GERMS OF DEATH

SUNY series in Contemporary French Thought
———————
David Pettigrew and François Raffoul, editors

GERMS OF DEATH
The Problem of Genesis in Jacques Derrida

MAURO SENATORE

Published by State University of New York Press, Albany

© 2018 State University of New York

All rights reserved

No part of this book may be used or reproduced in any manner whatsoever without written permission. No part of this book may be stored in a retrieval system or transmitted in any form or by any means including electronic, electrostatic, magnetic tape, mechanical, photocopying, recording, or otherwise without the prior permission in writing of the publisher.

For information, contact State University of New York Press, Albany, NY
www.sunypress.edu

Production, Jenn Bennett
Marketing, Fran Keneston

Library of Congress Cataloging-in-Publication Data

Names: Senatore, Mauro, author
Title: Germs of death : the problem of genesis in Jacques Derrida
Description: Albany : State University of New York Press, [2018] | Series: SUNY series in contemporary French thought | Includes bibliographical references and index.
Identifiers: ISBN 9781438468471 (hardcover   ) | ISBN 9781438468495 (ebook) | ISBN 9781438468488 (pbk.   )
Further information is available at the Library of Congress.

10 9 8 7 6 5 4 3 2 1

Living things die, and they do so simply because they carry the germ of death in themselves.

—Hegel, *Encyclopedia of Philosophical Sciences in Basic Outline, Part I: Science of Logic,* §92

Genealogy cannot begin with the father.

—Derrida, *Glas*

CONTENTS

ACKNOWLEDGMENTS	ix
PREFACE	xi

INTRODUCTION
The Inaugural Inscription	1
The Scene of Divine Creation	2
The Legacy of Husserl's *Origin*	8
The Most General Geneticism	14
The Generation of Consciousness	16
The Origin of Forms	20

CHAPTER 1
Platonism I: The Paternal Thesis	25
A Problem of Syntax	26
The Origin and Power of the Logos	31
The Textuality of Plato's Text	35
Autochthony	39
The Natural Tendency to Dissemination	42
The Science of the Disseminated Trace	45

CHAPTER 2
Platonism II: *Khōra*	49
The Earth of Fathers	51
The Boldness of Timaeus	57
The *Dynamis* of *Khōra*	61
The Concept of History	63

CHAPTER 3
Hegelianism I: Tropic Movements — 69
- The Philosophical Introjection of Ordinary Language — 70
- The Life of the Concept — 72
- The Hegelian Treatment of Equivocity — 76
- The Negation of Consciousness — 84
- The Two Deaths of the Metaphor — 87

CHAPTER 4
Hegelianism II: The Book of Life — 93
- The Systematic Figure of the Germ — 95
- The Tree of Life — 101
- The Circulation of Singular Germs — 106
- A Note on Classification — 112

CHAPTER 5
Hegelianism III: The Genetic Programme — 115
- This Is a Protocol — 116
- The Logic Text — 118
- The Two Deaths of the Preface — 120
- The Preface Is the Nature of the Logos — 128
- The Tain of the Mirror — 130
- A Non-*genetic* Thinking of Genesis — 135

Postscript — 139

Notes — 147

Bibliography — 177

Index — 183

ACKNOWLEDGMENTS

This book is the result of the Conicyt/Fondecyt Iniciación project n.11140145, hosted by the Instituto de Humanidades, Universidad Diego Portales (Santiago, Chile). My thanks are to all those without whom I would not have written it: Marguerite Derrida (for supporting this project), the IMEC: Institut Mémoires de l'édition contemporaine (for making its resources available), Francesco Vitale (for being a guide for my work), Juan Manuel Garrido, Rodolphe Gasché, David Johnson, Martin McQuillan, Michael Naas, and Eduardo Sabrovsky (for their generosity), Azeen, Matías, and Ronald (for their friendship), and my family, Paola, Nicola, Giulio, Catherine, and Edouard (to whom this book is dedicated).

PREFACE

In the early seventies, the biological paradigm of genetics, which had developed in France during the previous decade, began to tremble. According to this paradigm, biological heredity, transmitted from generation to generation and inscribed in the nucleic acid of a cell, constitutes the programme of development—the so-called genetic programme—of a living organism.[1]

In *L'Organisation biologique et la théorie de l'information* (*Biological Organization and Information Theory*, 1972), the young biophysicist Henri Atlan offers an initial formalization of the post-*genetic* understanding of life that he has improved throughout his subsequent work. In particular, in the last part of the aforementioned book, entitled *L'Organisation*, he calls into question the concept of the genetic programme. He argues that, since its primordial stage—that is, since the synthesis of proteins in the cell—the development of a living organism has consisted in the movement of self-organization through which a complex structure of life integrates the aleatory stimuli of the environment. Therefore, the synthesis of proteins would not amount to the mere execution of a set of pre-established instructions but to the process of disorganization and reorganization that derives from the irruption of the environment.[2] In the concluding section of the book, entitled "*Morts ou vifs?* [Dead or Alive]," Atlan summarizes his understanding of life through a formulation that is pregnant with implications for the biological thought to come:

> The dream of a cell is neither that of reproducing itself, nor of "enjoying" its metabolism, nor of assimilating, but, "like everyone [*comme tout le monde*]," that is, like every physical system in irreversible time—which is also the time of our representation—it is that of "resting" in the minimal state

of free energy and thus of dying.... Here, as well as in physical systems, the only project remains that of returning to equilibrium, namely, death.... The rest, that is, organization, growing, development, learning, and reproduction, are not of the order of the project but of the aleatory perturbations that succeed in contrasting it. Living organisms thus appear as systems that are enough *complex*, *redundant*, and *reliable* in reacting to the aleatory aggressions of the environment, so that the achievement of the state of balance, namely, death, is only possible through the detours that we agree on calling life. (Atlan 1972, 284, my translation)[3]

The analyses that follow are situated within these coordinates. They take up the thinking of genesis—namely, dissemination—that Jacques Derrida had elaborated in the sixties and the seventies, as a major contribution to theoretical debate surrounding post-*genetic* conceptions of life. This thinking demarcates itself from the philosophical understanding of genesis—what Derrida designates as the *logos spermatikos*—that undergirds the whole organization of knowledge and puts on the mask of preformationism in the life sciences. Furthermore, it demonstrates that genetics constitutes the latest feature of this preformationism. Therefore, Derrida's nonphilosophical and nonpreformationist understanding of genesis provides the conceptual framework for a post-*genetic* interrogation of life.[4] Later, Derrida takes this understanding as his point of departure for investigating ethico-political issues in the organization of living bodies.

This book elaborates a systematic examination of how Derrida develops the Husserlian concept of genesis through a critical engagement with Plato's and Hegel's legacies as well as with the biological thought of his time. This examination describes a trajectory throughout Derrida's early work that goes from his *Introduction* to Edmund Husserl's *Origin of Geometry* (1962) to *Glas* (1974). Within these limits, it draws together key moments in the demarcation of dissemination from the philosophy of genesis. The Introduction highlights the inaugural inscription of dissemination in the early essay "Force and Signification" (1963). In this text, Derrida broaches an articulation of the Husserlian concept of genesis, as the constitutive inscription of the ideal object, and of the Hegelian concept of the natural seed. He thus develops a new thinking of genesis in general (as a trace-seed) that is detached from the philosophical understanding of genesis (as a generation of consciousness). Chapters 1 and 2 offer an overall interpretation of Derrida's engagement with Plato's text, from "Plato's Pharmacy" (1968) to his edited and unedited writings on *khōra* (1970–1993). They point out that Derrida finds in this text the conflict between Platonism and dissemination. On the one hand, Platonism is interpreted as the thesis of the originary and nonmetaphorical relationship between the logos-*zōon* and its father-subject, which is embodied in the autochthonous community of Athens. On the other hand, dissemination stands for the thinking of the trace-seed as the general structure of genesis. Chapters 3 and 4 focus on the interpretation of the Hegelian text that Derrida develops from "Violence and Metaphysics" (1964) to *Glas* (1974). They explain that he understands this text as the book of life, in which the metaphorical exchange among ontological regions—for example, between nature and spirit—hinges on the presupposition that the truth of life is spiritual life or the life of the concept. Chapter 5 places dissemination in relation to the biological thought of the time by interpreting the preface to *Dissemination* (1972) as a critical response to *The Logic of Life* (1970), the masterwork of the French molecular biologist François Jacob. The chapter shows that Derrida's preface dissociates the understanding of genesis from the philosophical presuppositions that underlie Jacob's concept of the genetic programme. The book ends with a postscript that draws attention to a step forward taken by Derrida in his elaboration of the minimal conditions for genesis. Looking into significant moments in the seminar on the *GREPH*, delivered in 1974–1975, the postscript shows that the structure of genesis, the trace-seed, is attached to a drive for mastery and power.[5]

INTRODUCTION
The Inaugural Inscription

> The force of the work, the force of genius, the force, too, of that which engenders in general is precisely that which resists geometrical metaphorization and is the proper object of literary criticism.
>
> —Derrida, *Writing and Difference*, 23

> Preexistence excluded time and all notions of a history of life. Nature was entirely in the present, and the present itself flowed into the eternity of God. Nothing could have been more foreign to this period than the genetic character of Aristotelian form.
>
> —Roger, *The Life Sciences in Eighteenth-Century French Thought*, 311–12

In the following pages, I shed light on the inaugural inscription of dissemination—the thinking of genesis that is under exploration in this book—in Derrida's early essay "Force and Signification" (first published in 1963 and then included in *Writing and Difference*, 1967). I examine how Derrida welds his interpretation of the Husserlian notion of genesis to a modern concept of writing. On the one hand, we have the analysis of Husserl's regression to the transcendental origin of mathematics and of his elaboration of the constitutive traits of the first geometrical inscription. On the other hand, we have a rewriting of the Leibnizian scene of divine creation (that is, the transition of the best world possible into actuality) into a scene of the reproduction and alienation of sense, whose framework can be found in the account of the realization of species (namely, the seed) that Hegel gives in his philosophy of nature. Within these coordinates, I argue that Derrida conceives of the inscription-seed as the minimal structure of genesis in general, from biological to cultural genesis. This structure constitutes the element of a geneticism of sense in general—that is, of an analysis that accounts for the genesis of the discourse as well as of the living. Derrida's geneticism demarcates itself from the philosophical tradition of the *logos spermatikos*, whose understanding of any regional genesis is based on the paradigm of the generation of consciousness. According to Derrida, this tradition, which takes on the shape of preformationism in the life sciences and thus in the investigation of biological genesis, constitutes an imminent threat for the biology of his time (namely, molecular biology and genetics) as well as for literary criticism (for instance, the structuralism of Jean Rousset). In the final section of this introduction, I suggest reading Derrida's geneticism as a nonpreformationist response to the Aristotelian problem of the origin of forms that the historian of sciences Jacques Roger conjures up in his book on the life sciences in modern France, published in the same year as "Force and Signification."

THE SCENE OF DIVINE CREATION

In "Force and Signification," Derrida describes a decisive moment of literary creation, that of the transition of pure speech into inscription and thus into determination, as a natural and biological genesis, or, in the language of the Hegelian philosophy of nature—which, as we will see, underlies this description—as the realization of the species. I trace, throughout this early essay, Derrida's attempt to refer writing back to a process of engendering in general, through which sense (*le sens*) is constituted by differing from itself, and thus to prevent writing itself from being reappropriated within the self-reproduction and presence of a sense prescribed in the understanding of God.

Derrida accounts for the transition of pure speech into inscription through an explicit rewriting of the Leibnizian scene of creation. He conceives of speech and significations as Leibnizian essences (the possibles), to the extent that they are constituted by the claim for an inscription that brings them into existence (actuality):

Now, does not pure speech require inscription somewhat in the manner that the Leibnizian essence requires existence and pushes on toward the world, like power [*puissance*] toward the act? If the anguish [*angoisse*] of writing is not and must not be a *determined* pathos, it is because this anguish is not an empirical modification or state of the writer, but is the responsibility of *angustia*: the necessarily restricted passageway of speech against which all possible meanings [*significations*] push each other, preventing each other's emergence. Preventing, but calling upon each other, provoking each other too, unforeseeably and as if despite oneself, in a kind of autonomous overassemblage [*sur-compossibilité*] of meanings, a power [*puissance*] of pure equivocality [*equivocité*] that makes the creativity of the classical God appear all too poor. Speaking frightens me because, by never saying enough, I also say too much. And if the necessity of becoming breath or speech restricts meaning [*sens*]—and our responsibility for it—writing restricts and constrains speech further still. (Derrida 1978, 8–9)

In the scene of creation explained by Leibniz, essences and possibles struggle among each other to the extent that they all bear a claim for existence within themselves. In the short text entitled *On the Radical Origination of Things* (1697), Leibniz argues that the fact that "something exists rather than nothing" (1989, 487) entails that there is a certain tendency to existence in all possibles. This tendency (or urgency) has a different right (or force) according to the degree of reality (or perfection) of each possible. Hence, we learn that "this world should exist rather than another" (488). Leibniz writes:

To explain a little more distinctly, however, how temporal, contingent, or physical truths arise out of truths that are eternal and essential, or if you like, metaphysical, we should first acknowledge that from the very fact that something exists rather than nothing, there is a certain urgency [*exigentia*] toward existence in possible things or in possibility or essence itself—a pre-tension to exist, so to speak—and in a word, that essence in itself tends to exist. From this, it follows further that all possible things, or things expressing an essence or possible reality, tend toward existence with equal right in proportion to the quantity of essence or reality, or to the degree of perfection which they involve; for perfection is nothing but quantity of essence. (487)[1]

In the later *Principles of Nature and Grace* (1714, section 10), Leibniz points out that the struggle of the possibles laying claim to existence takes place in God's understanding and that "the result of all these claims must be the most perfect actual world which is possible"—that is, the one that "combines the greatest variety together with the greatest order" (639). This is the world chosen by God in the act of producing the universe, and it could not be otherwise given that it is supremely perfect.[2] If we step back to Derrida's version of the scene described by Leibniz, we observe that a claim to existence is not attached to each possible signification and thus to its degree of reality or perfection, but draws together, in an unpredictable and unmasterable fashion, possibles or essences that exclude each other—as Derrida puts it in the quoted passage, "a kind of autonomous overassemblage of meanings" and "a power of pure equivocality." In passing through the narrow extremity of a pen, speech and significations necessarily turn into something other than themselves—namely, an inscription. They are brought into existence or actuality by differing from themselves.[3]

A double, structural anxiety is called into play in the scene described by Derrida. First, there is "the anguish of writing," which has to do with the *angustia* and narrowness of the passageway and thus with an accumulation of claims that cannot be satisfied. This anxiety is linked to "responsibility" so long as writing responds to the demand for existence of speech and significations through a decision (Derrida 1978, 19).[4] But there is also another structural anxiety that affects what comes to light—namely, the breath that has been alienated and detached from its original source. This anxiety manifests when the breath withdraws. Derrida observes that there is "also, the anguish [*angoisse*] of a breath that cuts itself off in order to reenter itself, to aspirate itself and return to its original source" (383). I propose rereading this figure of double anxiety in light of a few remarks that Freud develops in *Inhibitions, Symptoms and Anxiety* (1926). In this text, Freud describes anxiety as "a state of unpleasure," provoked by an increase of excitation and accompanied by particular "acts of discharge" (1953–1974,

INTRODUCTION 5

132), such as the motor innervation of respiratory organs and of the heart. He identifies the traumatic experience that the state of anxiety comes to reproduce with the historical fact of birth. "In man," Freud explains, "birth provides a prototypic experience of this kind, and we are therefore inclined to regard anxiety-states as a reproduction of the trauma of birth" (132). A page below, he further elaborates upon this hypothesis in order to suggest that anxiety consists in the reproduction of an experience of danger and thus that birth, understood as the experience of separation from the mother, stands for the first state of anxiety. I quote a long passage from this elaboration:

> The earliest anxiety of all—the "primal anxiety" of birth—is brought about on the occasion of a separation from the mother. But a moment's reflection takes us beyond this question of loss of object. The reason why the infant in arms wants to perceive the presence of its mother is only because it already knows by experience that she satisfies all its needs without delay. The situation, then, which it regards as a "danger" and against which it wants to be safeguarded is that of non-satisfaction, of a *growing tension due to need*, against which it is helpless. I think that if we adopt this view all the facts fall into place. The situation of non-satisfaction in which the amounts of stimulation rise to an unpleasurable height without its being possible for them to be mastered psychically or discharged must for the infant be analogous to the experience of being born, must be a repetition of the situation of danger. What both situations have in common is the economic disturbance caused by an accumulation of amounts of stimulation, which require to be disposed of. It is this factor, then, which is the real essence of the "danger." (136).

Therefore, Freud understands anxiety as the experience of an increase in stimulation that cannot be mastered and whose historical origin is birth. Returning to Derrida's scene, I suggest that anxiety is a trait inherent in the act of writing as well as in the inscription, to the extent that the latter repeats the experience of birth. This provides us with a deeper insight about why for Derrida "the creativity of the classic God"—namely the God of Leibniz—is "still too poor."

In a note on the passage that I commented at the beginning of this section, Derrida formalizes the process of alienation as the very law of language: "To speak," he remarks, "is to know that thought must become alien to itself in order to be pronounced and to appear" (1978, 383). The authentic writer knows this law: "This is why," Derrida continues, "one senses the gesture of withdrawal, of retaking possession of the exhaled word, beneath the language of the authentic writer, the writer who wishes to maintain the greatest proximity to the origin of his act" (383). In this note, Derrida proposes understanding the concept of "originary language"—the language that remains close to the originary act of its generation and thus the language of the authentic writer—by referring to Feuerbach's account of "philosophical language" in the essay entitled *Towards a Critique of Hegel's Philosophy* (1839).[5] Feuerbach conceives of philosophical language by having recourse to the Hegelian notion of generation—namely, the realization of the species (*Gattung*). He explains that thought (truth, objective spirit, tradition, species, etc.) reappropriates or returns to itself through the narrow passageway of a mouth or a pen and thus through its necessary alienation (*Entäußerung*) into philosophical language. In other words, philosophical language is not absolutely detached from its origin, from thought itself, but remains in the proximity of it. This language is the element in which the species-thought reproduces itself and thus is fully present. Feuerbach writes:

> Philosophy emerges from mouth or pen only [the French translation quoted by Derrida has: *ne sort de la bouche ou de la plume que* . . .] in order to return immediately to its proper *source*; it does not speak for the pleasure of speaking—whence its antipathy for fine phrases—but in order not to speak, in order *to think*. To demonstrate is simply to show that what I say is true; simply to grasp once more the alienation (*Entäußerung*) of thought at the original source of thought. . . . Language is nothing other than the *realization of the species*, the mediation between the I and the thou which is to represent the unity of the species by means of the suppression (*Aufhebung*) of their individual isolation. (Derrida 1978, 383)

The implicit source of this passage is Hegel's *Philosophy of Nature* (1830) section 369, in which the "*realized (gewordene) genus* [*Gattung*]"—namely, the biological and natural seed, is determined as "the *negative identity* of the differentiated individuals" (1970, 414). For Hegel, this realization occurs only "in principle" insofar as, "on its *natural* side," the seed "is itself an immediate *singular,* destined to develop into the same natural individuality, into the same difference and perishable existence" (414). The species never reappropriates itself fully, as it occurs to the spirit, which, as Hegel remarks in the *Zusatz*, "exists in itself and for itself in its eternity" and thus amounts to an infinite and incorruptible presence (414). The seed brings about the irreversible destruction of its parents, through which the species comes back to itself—this is the meaning of *Aufhebung* in Hegel's text.[6] Therefore, the seed bears within itself the self-inequality of natural and biological life. In the passage quoted by Derrida, Feuerbach removes the distinction between the irreducible self-inequality of the species and the full presence of the spirit, between natural and spiritual life, and reinscribes the biological process of the realization of the species within the movement of the self-reproduction and incorruptibility of thought. *Aufhebung* turns out to account for this movement. Feuerbach supposes that the species-thought reproduces, conserves and circulates itself through the extremities of a mouth and a pen, and thus that it reappropriates or comes back to itself through these narrow passageways. To this extent, language would be the self-circulating and incorruptible seed of the spirit.[7] In recalling Feuerbach's concept of philosophical language, Derrida emphasizes the relationship between writing and the realization of the species, between inscription and the biological and natural seed. Throughout "Force and Signification," he seems to be concerned precisely with tracing writing back to Hegel's (and not to Feuerbach's) account of generation and *Aufhebung*.[8]

For Derrida, the God of Leibniz, conjured up as the classical paradigm of divine creativity ("God, the God of Leibniz . . ."), "did not know the anguish of the choice between various possibles" (Derrida 1978, 9), which constitutes the passion of writing as well as of engendering. First, there is no passageway from possibility to actuality, nor from essence to existence, as it occurs in the scene described by Derrida. The God of Leibniz "conceived possible choices in action," Derrida writes, "and disposed of them as such in his Understanding or Logos" (9). Second, the passageway at stake in the Leibnizian scene is narrow as it consists in the will of God himself, which lets only the best go through and come into existence. As Derrida remarks, "In any event, the narrowness of a passageway that is *Will* favors the 'best' choice" (9). The implications of this classical paradigm are that each existence is reappropriated in the self-circulation and full presence of the same and identical Book and thus no engendering, no realization of the species, no inscription-seed take place.[9] "And each existence," Derrida observes, "continues to 'express' the totality of the Universe. Therefore, there is no tragedy of the book. There is only one Book, and this same Book is distributed throughout all books" (9).[10] He recalls *Theodicy* sections 414–16, in which Pallas, Jupiter's daughter, guides Theodorus through the Palace of Fates. As she explains to him, the palace guards the "representations of all that is possible" (Leibniz 1952, 375), which God made into worlds before choosing the best. Therefore, Theodorus is granted the privilege to see all possible worlds and how each of them comprises all existences. Pallas says:

> I have only to speak, and we shall see a whole world that my father might have produced, wherein will be represented anything that can be asked of him; and in this way one may know also what would happen if any particular possibility should attain unto existence. And whenever the conditions are not determinate enough, there will be as many such worlds differing from one another as one shall wish, which will answer differently the same question, in as many ways as possible. (375)

Derrida emphasizes the moment in which Theodorus is led into a hall—one of the worlds that Jupiter might have produced—and asks about a volume of writings. Pallas explains that it is the book of fates of this world and that he can see there the history of Sextus in all details by simply pressing his finger on a line. "He obeyed," Leibniz writes, "and he saw coming into view all the characteristics of a portion of the life of that Sextus" (Derrida 1978, 376–77). As we know, this is possible only because Sextus's life expresses the book of fates of one of the possible worlds and, thus, because this book circulates and returns to itself while being alienated into that life.

THE LEGACY OF HUSSERL'S *ORIGIN*

Derrida traces the nonclassical concept of writing-engendering back to the legacy of Husserl's *Origin of Geometry*, which he had highlighted in the *Introduction* to his translation of the text (1962).[11] In the following passage, the *Origin of Geometry* is convoked again in order to account for modernity—that is, for a certain break with the classical and Leibnizian concept of creation in which existences and geneses are neutralized.

> It [writing] is also to be incapable of making meaning absolutely precede writing: it is thus to lower meaning while simultaneously elevating inscription . . . To write is to know that what has not yet been produced within literality has no other dwelling place, does not await us as prescription in some *topos ouranios*, or some divine understanding. Meaning must await being said or written in order to inhabit itself, and in order to become, by differing from itself, what it is: meaning. This is what Husserl teaches us to think in *The Origin of Geometry*. (Derrida 1978, 11)

In *Introduction* sections 6–7, Derrida explains how the geometric ideality ("just like that of all sciences," [Derrida 1989, 76]) passes from an originally intrapersonal emergence to its ideal objectivity through the mediation of language.[12] In section 6, he points out that here Husserl, who "seems redescending toward language" and, therefore, more generally, toward culture and history, "does exactly the opposite" (77). "The return to *language*," Derrida continues, "brings to its final completion the purpose of the reduction itself" (77), by liberating the ideality from the psychological life of a factic individual community ("the inventor's head," [78]) in which it has emerged first, and by letting it be what it is (namely, other than itself). Therefore, language is "constitutive" with respect to sense, which, otherwise, would remain "ineffable and solitary" (78).[13] However, still another reduction is required to accomplish the passage of sense from the originally psychological formation to its historical constitution. In section 7, Derrida observes, in the wake of Husserl, that only writing permits the full accomplishment of the ideal objectivity of ideality by unbinding the latter from an actual subjectivity in general, that of the inventor as well as of the community of his fellows, and thus by granting the ideality's traditionalization—that is, the possibility of its omnitemporal and omnispatial reactivation.[14] Returning to "Force and Signification," it seems that, by drawing together the legacy of the *Origin of Geometry* and the above discussed rewriting of the Leibnizian scene of creation, Derrida reinscribes his recent interpretation of Husserl into a more general formulation of the problem of genesis. Husserl's concept of writing as well as of the historical genesis of sense comes back here to account for the nonclassical concept of writing-engendering.

However, it is in *Introduction* section 2, I argue, that Derrida develops more explicitly the Husserlian concept of genesis that "Force and Signification" retrieves as the legacy of the *Origin*. The section unfolds a comparative analysis of Kant's and Husserl's regression toward a nonempirical and transcendental origin of geometry.[15] Derrida draws attention to a passage from the Preface to the *Critique of Pure Reason*, in which Kant identifies the origin of mathematics as the moment when the individual's idea opens up the tradition of mathematics and thus the horizon of its possibility and future developments. He paraphrases the Preface as follows:

> "The history of this revolution," attributed to the "happy thought of a single man" in "an experiment from which the path that had to be taken must no longer be missed and from which the sure way of science was *opened* and *prescribed* (*eingeschlagen und vorgezeichnet war*) for all times and in endless expansion (*für alle Zeiten und in unendlich Weiten*), was more "decisive" than the empirical discovery "of the path around the famous Cape [of Good Hope]." (39)

As emphasized in this passage, Kant conceives of the nonempirical and transcendental origin of mathematics as a pre-*scription*. Therefore, he seems to share with Husserl the concept of "the original genesis of a truth, whose birth (or birth certificate) [*act de naissance*] inscribes and prescribes omnitemporality and universality" (39). However, as Derrida points out, the Kantian genesis—what Derrida calls the "inaugural mutation"—is not creative: it is not so much a revolution as a "revelation" (39). "It is not *produced* [my emphasis] by him," the first geometer, as Derrida writes, yet "it is understood under a dative category, and the activity of the geometer to which the 'happy thought' occurred is only the empirical unfolding [*déploiement*] of a profound reception" (39–40). The legacy of the Husserlian notion of genesis consists precisely in the productive, constitutive, or creative trait of the first geometrical inscription.[16] The beautiful passage that follows uncovers the minimal condition for genesis—namely, for a genesis that is not reappropriated by the self-circulation of sense but brings the latter to existence and makes it differ from itself:

> Undoubtedly, Husserl's production (*Leistung*) also involves a stratum of receptive intuition. But what matters here is that this Husserlian intuition, as it concerns the ideal objects of mathematics, is absolutely constitutive and creative: the objects or objectivities that it intends did *not* exist *before* it; and this "*before*" of the ideal objectivity marks more than the chronological eve of a fact: it marks a transcendental prehistory. In the Kantian revelation, on the contrary, the first geometer merely becomes conscious that it suffices for his mathematical activity to remain within a concept that it *already possesses*. The "construction" to which he gives himself, then, is only the explication of an already constituted concept that he encounters, as it were, in himself. (Derrida 1989, 40)[17]

The temptation to compare this passage to the later rewriting of the Leibnizian scene of creation is difficult to resist. The layer of ideal objects, which stands before the constitutive genesis, that is, before the birth of geometry, and accounts for "a transcendental prehistory," may be thought as an autonomous overassemblage of possibles or essences, as a pure equivocity that claims for existence. Finally, having recourse to a language that comes back in "Force and Signification," Derrida remarks that, for Kant, there is no subject that takes on the responsibility of the constitutive production and thus of carrying sense out of itself without any assurance of self-return or self-reappropriation. "Strictly speaking," he writes, "the reduction [the fulfilment of ideal objectivity] is not for or by a subject who makes himself *responsible* for it in a transcendental *adventure* [my emphasis]" (41–42).

The nonclassical writing evoked in "Force and Signification" is inaugural as well as the historical genesis of geometry in the *Origin*. For this reason, Derrida writes, "It is dangerous and anguishing" (1978, 11). It makes sense be what it is by becoming other than itself, and thus it starts the movement of anticipation and delay of sense. Derrida continues: "Writing . . . does not know where it is going, no knowledge can keep it from the essential precipitation toward the meaning that it constitutes and that is, primarily, its future" (11). This movement can be imagined as the text made of those historical geneses that Husserl evokes in the *Origin*, the text into which a book without genesis, such as the Leibnizian Book, has necessarily been reinscribed. A few pages later, Derrida proposes to designate this text by "volume," and he describes it as "the infinite implication" and "the indefinite referral [*renvoi*] of signifier to signifier" (29)—that is, of genesis to genesis. He supposes that the impulse or motivation of this movement-text, its force, is pure equivocity, or the transcendental prehistory of idealities, essences, possibles, and thus sense, which claims for coming to existence through the narrow medium of writing. Derrida designates the text of the historical geneses of sense in general as this movement of anticipation and delay of sense. "Is it by chance," Derrida wonders, "that its force [of the volume] is a certain pure and infinite equivocality which gives signified meaning no respite, no rest, but engages it in its own *economy* so that it always signifies again and differs?" (29).

A few years later, in a passage on the left-hand column of *Glas* (1974), Derrida finds in Hegel's inheritance of Schelling's concept of *Potenz* a source of the general economy of sense sketched out in "Force and Signification." This passage from *Glas* offers a noteworthy explanation of the movement of anticipation and delay, which, in this case, involves the idealistic absolute rather than sense. Derrida begins by recalling that the term *Potenz* can be read as a moment of the Hegelian concept of moment. It occurs with a certain frequency in *On the Scientific Ways of Treating Natural Law* (known as *Natural Law*, 1802–1803) before becoming a key feature of the Jena philosophy of spirit. This concept, Derrida remarks, constitutes a legacy of Schelling's philosophy of nature, where it stands for the totality into which the absolute leaps when it goes out of itself. In accounting for Schelling's concept of *Potenz*, Derrida (1986) describes the economy through which the absolute comes to manifestation by becoming other than itself:

> The *Ideas for a Philosophy of Nature* describe the absolute's going outside itself into nature, as nature, according to the ternary rhythm of "powers" [*puissances*]. These "powers" are at once a *dynamis* and an *energeia*, a virtuality and an act, a completed totality on which, so to speak, is hung the totality to come. . . . The absolute goes out of itself into the finite, penetrates the finite with its infinity in order to make the finite come back to it. It absorbs it, resorbs it after having

entered it. This movement of effusion/resorption manifests the absolute differentiating itself, going out of the night of its essence and appearing in the daylight. (105)

The same schema is found in Schelling's *The Ages of the World* (1815), where "the abyssal absolute" is precisely "powerless" (*Potenzlos*). Hegel puts this schema to work in *Natural Law*, in which each determinate totality of the text is a *Potenz*, and thus the text develops through a rhythm of *Potenzen*. Derrida focuses on the last pages of the text, where Hegel reflects upon the life and death of the ethical totality in general. He focuses on the "archeo-teleological movement" in which the absolute totality simultaneously anticipates and defers itself through finite totalities.

> In each particular totality, as such, the absolute totality *comes to a halt* [*s'arrête*], *stops itself*, stops its necessity. The particular totality then takes, as a part, a certain independence, a certain subsistence. To come to a halt, to stop itself, is here *sich hemmen*. *Hemmen* is often translated by inhibit, suppress. The infinite totality inhibits itself in the *Potenz*. This totality limits itself, gives itself a form, goes out from a certain *apeiron*, suspends itself, puts an end to itself, but the delay it thus takes on itself (*hemmen* also signifies delay, defer) is the positive condition of its appearing [*apparaître*], of its glory. Without the delay, without the suspensive and inhibiting constriction, the absolute would not manifest itself. So the delay is also an advance, progress, an anticipation, an encroachment on the absolute unfolding of the absolute. (Derrida 1986, 106)

The archeo-teleological movement that regulates this absolute unfolding constitutes the chain or text of finite totalities into which the absolute is constrained. Its necessity, Derrida concludes paraphrasing Hegel, is "the conflict of forces," according to which the most powerful *Potenz* imposes itself by suppressing the others, at the same time as it lays the condition for a new *Potenz*.[18] In the text discussed by Derrida, Hegel points out that the notion of *Potenz* is the particular with which philosophy concerns itself: not the particular as such but as the totality into which the absolute passes when it goes out of itself.

The target of this philosophical understanding is formalism, which regards the particular as contingent and dead. "The absolute totality, as a necessity," Hegel (1999) writes, "confines itself within each of its potentialities [*Potenzen*], and produces itself as a totality on this basis. It there recapitulates the [development of] preceding potentialities, as well as anticipating [that of] those still to follow" (178).

THE MOST GENERAL GENETICISM

The birth of language occurs when the latter is a constitutive and historical genesis that interrupts the supposed self-circulation of the essences prescribed in the understanding of God and takes place in the movement of the precipitation and delay of sense. Language is by definition defunct, deprived of signalizing function, and thus it is a constitutive and historical genesis, like the seed of the Hegelian philosophy of nature, which carries the death of its genitors.[19] Derrida (1978) highlights the bond of language and birth, writing and seed, as follows:

> It is when that which is written is *deceased* [the French *defunt* plays with *function*: it refers to the liberation from the signalizing function and, thus, to pure functioning] as a sign-signal that it is born as language; for then it says what is, thereby referring only to itself, a sign without signification, a game or pure functioning, since it ceased to be *utilized* as natural, biological, or technical information, or as the transition from one existent to another, from a signifier to a signified. (13)

Derrida conceives of writing as the structure of genesis in general, from biological to cultural genesis. From the biological perspective, dissociating writing from information and thus rethinking the former in light of the indefinite implication of signifiers and the general economy of sense constitute an enormous task. Here Derrida seems to reject the understanding of biological heredity—which is constitutive of all living beings—as information and, thus, as the self-circulation of the Leibnizian Book. Furthermore, he seems to suggest that heredity is a historical genesis, a genetic inscription. Only a couple of years later, in *Of Grammatology* (whose first part is originally published in

1964–1965), he finds a scientific support to his speculative hypothesis in the contemporary discovery of the nucleic acid (DNA), in which biological heredity is inscribed (genes), and in its conceptualization through the cybernetic notion of pro-*gramme*.[20] It is also worth noting that, a year before "Force and Signification," in 1962, in the Royaumont colloquium on "the concept of information in contemporary science," the molecular biologist André Lwoff presented a paper entitled "Le Concept d'information dans la biologie moleculaire" ("The Concept of Information in Molecular Biology").[21] The paper describes biological heredity—that is, the inherited genes that regulate the development and differentiation of life in the cell (the synthesis of proteins) as a spatio-temporal sequence and, ultimately, as a text. The concept of information is at the center of Lwoff's paper, who designates the latter as "what determines life" (Cahiers de Royaumont 1965, 173, my translation). Lwoff points out that biological information is something "material" insofar as it amounts to a spatio-temporal sequence or a text. "The foundation of biological order," he writes, "is therefore, a determinate sequence, a sequence in the space as well as in time" (180). In so doing, instead of dissociating information from writing, he ends up identifying one with the other. There is no evidence that, in the quoted passage, Derrida refers to Lwoff, but we may suppose that he is advising any biological theorists of heredity against falling back into a classical concept of genesis and thus, as we see below, against affirming once more the biological doctrine of preformationism.

In the last pages of part I of "Force and Signification," Derrida offers an explicit formulation of the geneticism [*génétique*] of sense that he has aimed to formalize by rewriting the classical and Leibnizian scene of divine creation. He argues that the vulnerability of structuralism in general and of Jean Rousset's criticism, in particular, to whose exploration his essay is explicitly devoted, consists in neutralizing the internal historicity of a work (its relation to a subjective origin such as force, genius, sense, etc.) and a certain inability to see it in the present (duration). By taking up the form of a work as its sense, structuralism does with works what Leibniz does with existences: it neutralizes any genesis insofar as the latter expresses or transports the information prescribed in the Book. Derrida (1978) notes:

> It is true that in some places the form of the work, or the form as the work, is treated *as if* it had no origin. . . . By

keeping to the legitimate intention of protecting the *internal* truth and meaning of the work from historicism, biographism or psychologism (which, moreover, always lurk near the expression "mental universe"), one risks losing any attentiveness to the internal historicity of the work itself, in its relationship to a subjective origin that is not simply psychological or mental . . . one risks overlooking another history, more difficult to conceive: the history of the meaning of the work itself, of its *operation*. This history of the work is not only its *past*, the eve or the sleep in which it precedes itself in an author's intentions, but is also the impossibility of its ever being *present*, of its ever being summarized by some absolute simultaneity or instantaneousness. This is why, as we will verify, there is no *space* of the work, if by space we mean *presence* and *synopsis*. (15)

The alternative to this structuralist conception of work, Derrida maintains, is "an internal geneticism, in which value and meaning are reconstituted and reawakened in their proper historicity and temporality" (15). This geneticism should be thought in light of the examined attempt to trace writing and cultural genesis back to biological and natural engendering. It would constitute the most general geneticism, that of the genius, of the force of engendering in general, of *physis*. Furthermore, as suggested above, it would be articulated with the biology of the time, namely, genetics, insofar as it finds in writing the minimal condition for the genesis of history and sense.[22]

THE GENERATION OF CONSCIOUSNESS

In "Force and Signification" part 2, which engages in the reading of Rousset's criticism, Derrida has recourse to the concept of the *logos spermatikos* as the paradigm of classical creation. He focuses on Rousset's reading of Proust, in which the structuralist neutralization of the force and duration of a work takes on the form of the biological theory of preformationism. He begins by discussing the specific moment in which Rousset describes Proust's aesthetics of the novel as "the phenomenology of spirit" (25). Derrida observes that "the implication of the end in the beginning, the strange relationship between the subject who writes a book and the subject of this book," as they are emphasized in Rousset's reading of Proust, "recall the style of becoming and the dialectic of the 'we'" (25) in the Hegelian *Phenomenology of Spirit* (1807). Rousset explains:

> One can discern still more reasons for the importance attached by Proust to this circular form of a novel whose end returns to its beginning. In the final pages one sees the hero and the narrator unite too, after a long march during which each sought after the other, sometimes very close to each other, sometimes very far apart; they coincide at the moment of resolution, which is the instant when the hero becomes the narrator, that is, the author of his own history. The narrator is the hero revealed to himself, is the person that the hero, throughout his history, desires to be but never can be; he now takes the place of this hero and will be able to set himself to the task of edifying the work which has ended, and first to the task of writing *Combray*, which is the origin of the narrator as well as of the hero. The end of the book makes its existence possible and comprehensible. The novel is conceived such that its end *engenders* [my emphasis] its beginning. (Derrida 1978, 26)

A few pages earlier, resorting to the metaphor of the development of a germ, Rousset (1962) observes that Proust's novel is more than this development: it is "the history of a spirit and of its salvation" (139, my translation). Therefore, Rousset interprets the phenomenology of spirit as the movement through which the spirit reproduces itself. "As in Hegel," Derrida points out, "the philosophical, critical, reflective consciousness is not only contained in the scrutiny given to the operations and works of history. What is first in question is the history of this consciousness *itself* [son *histoire*]" (1978, 39).[23] In the final pages of the "Introduction" to the *Phenomenology*, Hegel illustrates the phenomenology of spirit—namely, the style of becoming and the dialectics of the *we*—that Rousset (and Derrida) finds at work in Proust's aesthetics of the novel. Hegel remarks that, in the presentation (*Darstellung*) of the course of experience that makes up the phenomenology of spirit, we must distinguish between the ordinary consciousness of experience—that is, of the transition from one object of knowledge to another—as the pure apprehension of what we come upon by chance, and *our* perspective, which looks at the succession of experiences as the scientific progression of the ordinary consciousness toward the philosophical consciousness represented by *us*.[24]

> This exposition of the course of experience contains a moment in virtue of which it does not seem to agree with what is ordinarily understood by experience. This is the moment of transition from the first object and the knowledge of it, to the other object, which experience is said to be about. Our account implied that our knowledge of the first object, or the being for-consciousness of the first in-itself, itself becomes the second object. It usually seems to be the case, on the contrary, that our experience of the untruth of our first notion comes by way of a second object which we come upon by chance and externally, so that our part in all this is simply the pure *apprehension* of what is in and for itself. From the present viewpoint, however, the new object shows itself to have come about through a *reversal of consciousness itself.* This way of looking at the matter is something contributed by *us,* by means of which the succession of experiences through which consciousness passes is raised into a scientific progression—but it is not known to the consciousness that we are observing. (Hegel 1977, 55)

The "necessity" of the development and thus of "the origination of the new object" is what "proceeds for us, as it were, behind the back of consciousness." Therefore, there is a *we* that is "non-present" to the consciousness immersed in its experience and that grasps experience itself "as movement and a process of becoming" (56). Now, phenomenic consciousness comes to coincide with this *we* in the final point of Hegel's exposition, in which it merges with the science of the experience of consciousness and thus with *our* absolute knowledge.

> In pressing forward to its true existence, consciousness will arrive at a point at which it gets rid of its semblance of being burdened with something alien, with what is only for it, and some sort of 'other,' at a point where appearance

becomes identical with essence, so that its exposition will coincide at just this point with the authentic Science of Spirit. And finally, when consciousness itself grasps its own essence, it will signify the nature of absolute knowledge itself. (56–57)

Rousset refers to this Hegelian *point* in his passage on Proust, when he speaks about the "moment of resolution" in which "the hero becomes the narrator, that is, the author of his own history." For Derrida, this understanding of the literary work consists in the application of the biological theory of preformationism to human consciousness. The theory of preformationism, he explains, conceives of the living adult as the unfolding of the hereditary characters folded and thus preformed in the germ.[25] Therefore, it formalizes the self-reproduction of the prescribed book of life. However, according to Derrida, this theory consists in the antropomorphization of nature as engendering in general (*physis*), insofar as it applies the engendering of human consciousness—namely, self-reproduction or the *logos spermatikos*—to nature itself. As Derrida puts it, preformationism "attributes something more than finality to natural life—providence in action and art conscious of its works" (1978, 26). But "when the artist is a man, and when it is consciousness that engenders," he continues, "preformationism no longer makes us smile. *Logos spermatikos* is in its proper element, is no longer an export, for it is an anthropomorphic concept" (26–27). This anthropomorphic concept of engendering and nature converges with the classical and Leibnizian concept of creation, which neutralizes any natural as well as historical geneses and thus the general economy of the anticipation and delay of sense. Derrida opposes his geneticism of sense in general to this concept. The metaphysics that underlies structuralist preformationism and, more generally, the concept of the *logos spermatikos* is summarized in the aforementioned pages of Leibniz's *Theodicy*. The Book actualized in the understanding of God conserves itself and thus it is incorruptible and accessible in the Palace of Fates to a privileged visitor such as Theodorus. For Derrida (1978):

> It is demonstrable that what is in question [in the aesthetics of Proust] is the metaphysics implicit in all structuralism, or in every structuralist proposition. In particular, a structuralist reading, by its own activity, always presupposes and appeals

to the theological simultaneity of the book, and considers itself deprived of the essential when this simultaneity is not accessible. (28)

THE ORIGIN OF FORMS

In conclusion, I draw attention to a book published in the same year as "Force and Signification," which helps us understand the stakes of Derrida's commitment to demarcating his general geneticism from the tradition of the *logos spermatikos*. This book is the published version of the doctoral thesis of the French historian of sciences Jacques Roger, entitled *The Life Sciences in Eighteenth-Century French Thought (The Generation of Animals from Descartes to the Encyclopedia)*.

In chapter 6, dedicated to "Preexistence of Germs," the historian explores how preformationism develops into the theory of preexistence in seventeenth and eighteenth century Europe. According to preformationism, Roger (1999) explains, true generation (namely, "the actual formation of the living being") consists in the explication of the living being, which is contained as entirely formed (as a "germ") in the vegetal seed or the animal semen. "Embryonic development," he writes, "was thus no longer formation but simply enlargement of already existent components" (259). The doctrine of preexistence shares with preformationism the preexistence of the living being completely formed in the seed or semen, but adds that the germ "had been created by God at the beginning of the world and had been preserved since then until the moment of its development" (260). Roger observes that the two theories aim to respond to the same philosophical problem, that of the generation of living beings as the origin of forms. This problem became urgent after the decline of Aristotelianism, which explained the formation of forms through the relationship between potentiality and actuality. "The form of a being," Roger summarizes, "appears when that being, having existed potentially in matter, is realized in actuality" (261). To illustrate the context of preformationism and preexistence, Roger recalls Jean Fernel's argument against the Aristotelian explanation of the origin of forms. According to Fernel, a being is not transformed through the transition from potentiality to actuality and, therefore, a form cannot come to existence out of potentiality. Rather, it must have already been there, in the embryo. "Then Fernel has only one way of explaining the origin of forms," Roger concludes: "they must descend from heaven" (261); in the lexicon of Platonism, they must emanate from eternal ideas and forms. Ultimately, the genesis of natural and living beings is neutralized by the self-reproduction of the Book of God: "Nothing appeared in it that did not come from the original creation of all things" (261).

Roger suggests that this anti-Aristotelian thought of generation betrays the influence of the Stoic theory of seminal reasons (namely, *logoi spermatikoi*) as Augustine elaborates it in *De Genesi ad litteram* (book 10). By attributing to germs a principle of activity that enables them to develop in the right moment, this theory allowed Augustine to explain how new beings are still being produced in nature although God has already created everything. Therefore, no constitutive trait remains in natural genesis: as Roger remarks through Etienne Gilson's words, the doctrine of seminal reasons "eliminates . . . all hint of creative efficacy in the activity of man and other created beings" (264). By referring to the expression from *Ecclesiastes* "Deus creavit Omnia simul," Augustine acknowledges that the neutralization of genesis occurs in the simultaneity of creation.[26] Roger focuses on the formalization of the Augustinian doctrine of the preexistence of germs developed by Malebranche in *The Search after Truth* section I.2, where Malebranche takes into account the error of senses with regard to absolute extension. He pushes the implications of this theory to their limits by linking the idea of preformation to that of the encasement related to preexistence. He explains that, even if we cannot perceive preformation by senses and we can only suppose it through optical instruments, "it is reasonable believe" that "all trees are in the seeds of their seeds in miniature" and, therefore, that "there is an infinite number of trees in a single seed, since it contains not only the tree of which it is seed but also a great number of other seeds that might contain other trees and other seeds . . . and so to infinity" (1980, 26–27). These implications result "strange and incongruous" Malebranche continues, "only to those who measure the marvels of God's infinite power by the ideas of sense and imagination" (27). Conversely, for Derrida, preformationism can be understood only as the exportation of the generation of consciousness to nature, and thus it has always already escaped the question about the necessity of taking recourse to the concept of natural genesis or generation in order to explain the conservation and presence of consciousness. Malebranche's conclusion is that the role of nature, as the genesis and existence of living beings, is reduced to the unfolding of tiny trees and thus to the difference between each tiny tree and its explication.

The debate surrounding the Aristotelian problem of the origin of forms is not at all alien to "Force and Signification," as it might appear at first glance. Derrida himself inscribes his geneticism into this debate by demarcating the singularity of his position. He suggests that the metaphysical foundations of Rousset's structuralism are to be found in a specific moment of Aristotle's reflection on potentiality and actuality. According to Derrida, Rousset's understanding of movement in literature hinges on the definition of physical movement given by Aristotle in *Physics* 3: "The entelechy of potentiality, as such, is motion" (1.201a10–201a14).[27] The Aristotelian passage is paraphrased as follows: "Transition [*passage*] to the act, which itself is the repose of the desired form" (Derrida 1978, 24). Therefore, on Derrida's reading, movement—and thus inscription, genesis, and so forth—has already been reappropriated within the theological and teleological perspective of the desired form. In the aforementioned text from *Physics*, Aristotle explains that "motion occurs just when the fulfilment itself occurs, and neither before nor after." For instance, he notes, "the actuality of the buildable as buildable is the process of building." This actuality is not the one of the house, because "when there is a house, the buildable is no longer there" (1.201b7–201b15). However, we cannot think of motion as dissociated from the entelechy of the process, even if this is not the same as the actuality of the result, in which there is no longer any trace of motion itself. In *Metaphysics* 9, Aristotle remarks explicitly that we would be unable to conceive of movement and becoming without having recourse to the concepts of potentiality and actuality.

> Again, if that which is deprived of potentiality is incapable, that which is not happening will be incapable of happening; but he who says of that which is incapable of happening that it is or will be will say what is untrue; for this is what incapacity meant. Therefore these views do away with both movement and becoming. For that which stands will always stand, and that which sits will always sit; if it is sitting it will

not get up; for that which cannot get up will be incapable of getting up. But we cannot say this, so that evidently potentiality and actuality are different; but these views make potentiality and actuality the same, so that it is no small thing they are seeking to annihilate. (1047a11–1047a20)

Does this passage ground what Derrida understands as structuralism in general—namely, the neutralization of genesis and thus of the historicity and duration of movement and becoming?[28] Through his interpretation of structuralism, Derrida distances himself from preformationism as well as from what we may designate as the Aristotelian response to the philosophical problem of the origin of forms, which, in fact, underlies preformationism itself. Ultimately, he proposes an original geneticism that takes the Hegelian concept of generation—namely, the realization of the species—as the minimal condition for biological as well as cultural genesis.

In the analyses that compose this book, I explain how Derrida develops the preliminary inscription of dissemination that I have traced here, through the readings of Plato's and Hegel's texts that he undertakes subsequently. As we will see, the stakes of these readings are essentially interwoven with the task of dissociating dissemination from what Derrida designates as the inaugural version of the *logos spermatikos* (Platonism) and its most explicit elaboration (the Hegelian book of life).

1

PLATONISM I

The Paternal Thesis

> How can we set off in search of a different guard, if the pharmaceutical "system" contains not only, in a single stranglehold, the scene in the *Phaedrus*, the scene in the *Republic*, the scene in the *Sophist*, and the dialectics, logic, and mythology of Plato, but also, it seems, certain non-Greek structures of mythology? And if it is not certain that there are such things as non-Greek "mythologies"—the opposition *mythos/logos* being only authorized *following* Plato—into what general, unnamable necessity are we thrown? In other words, what does Platonism signify as repetition?
>
> —Derrida, *Dissemination*, 167–68

Throughout his reading of Plato's text, Derrida demarcates dissemination from the understanding of genesis that he calls "Platonism." I start my exploration of this reading by focusing on the earliest moment of it, the long essay "Plato's Pharmacy," first published in *Tel Quel* (1968) and then included in *Dissemination* (1972). In this essay, Derrida describes Platonism as the thesis that the living *logos*, assisted by its father and determined by the traits of the noble birth and the body proper, is the element of all regional discourses, from linguistics to zoology, from cosmology to politics. He understands this thesis as the myth itself, the story that the logos tells (a mytho-logy) about its origin—that is, its originary and nonmetaphorical relation to its father.

Platonism tends to annihilate what Derrida identifies as its anagrammatic structure—namely, the site of the concatenations of forms, of the tropic and syntactical movements, which precede and render possible the concatenation or movement of Platonism itself, as well as of philosophy in general. I examine "Plato's Pharmacy" by taking as my point of departure session 2 of the recently edited course on *Heidegger: The Question of Being and History* (1964–1965). My argument is that the later essay can be reread as an elaboration of Derrida's earlier analysis of Heidegger's insight that philosophy demarcates itself from mythology for the first time in Plato. Therefore, a path between two notions of grammar, or syntax, awaits us. On the one hand, we have Heidegger's search for a grammar for the destruction of the history of ontology and the demarcation of philosophy from mythology. On the other hand, we have grammar as the science of the concatenations of elements, invented by the Egyptian god Theuth, which, for Derrida, constitutes the science of the origin of the world, of the living as well as of the logos, of the disseminated trace.

A PROBLEM OF SYNTAX

In session 2 of *Heidegger: The Question of Being and History*, Derrida focuses on the problem of language concerning the destruction announced by Heidegger—that is, the destruction of the history of ontology as "a covering-over or a dissimulation of the authentic question of Being, under not ontological but ontic sedimentations" (Derrida 2016, 1).[1] In so doing, he brings to the fore an issue that Heidegger confines to a marginal place. As he acknowledges, the question "is posed in an added remark, which is a little surprising and, if I have forced Heidegger's thinking, it is by placing this added remark in the foreground" (25). This remark is included in the final paragraph of the introduction to *Being and Time*, dedicated to the "Exposition of the Question of the Meaning of Being." Heidegger presents this paragraph as a supplementary remark on the style of his subsequent analyses, which he demands the reader to measure against the task that is being undertaken in the book. I propose inverting the movement of Derrida's text by starting with the Heidegger passage that Derrida quotes and, from this, going back to the latter's formulation of the problem of language. Heidegger's remark reads:

With regard to the awkwardness and "inelegance" of expression in the following analyses, we may remark that this is one thing to report narratively about *beings* another to grasp beings in their *being*. For the latter task not only most of the words are lacking but above all the "grammar." If we may allude to earlier and in their own right altogether incomparable researches on the analysis of being, then we should compare the ontological sections in Plato's *Parmenides* or the fourth chapter of the seventh book of Aristotle's *Metaphysics* with a narrative passage from Thucydides. Then we can see the stunning character of the formulations with which their philosophers challenged the Greeks. Since our powers are essentially inferior, also since the area of being to be disclosed ontologically is far more difficult than that presented to the Greeks, the complexity of our concept-formation and the severity of our expression will increase. (1996, 34)[2]

In the pages that precede the quotation from Heidegger's text, Derrida anticipates the problem of the language of destruction by highlighting the feature of the forms of concatenation (*enchainement*). "Whence are we to draw the concepts, the terms, the forms of linking [*enchainement*] necessary for the discourse of Destruction, for the destructive discourse?" (Derrida 2016, 23–24), he wonders. A few paragraphs later, he develops this reference to the forms of concatenation of language and discourse by reformulating the problem of the language of destruction as mainly a question of syntax, where syntax is implicitly understood to designate the science of the concatenation of concepts and words. The problem of language, he notes, "is not only a problem of philosophical lexicology, but it is a problem of syntax which concerns the forms of linkage [*enchainement*] of concepts" (25). Here Derrida sheds light on Heidegger's introductory remark that the task of the subsequent analyses is jeopardized by a lack of syntax. In the following pages, he sets out a careful examination of this remark that takes his exposition beyond the boundaries of Heidegger's text, toward a seminal reading of Plato's *Timaeus*. This examination is developed under the heading "*ontic metaphor*" (26), which seems to resonate with the ontic and not ontological sediments that dissimulate the question of Being. Derrida reformulates, once again, the problem of language by linking it to Heidegger's self-inhibition of narrative. "The language difficulty," he explains, "hangs, then . . . on the fact that for the first time we are going to forbid ourselves resolutely and absolutely from 'telling stories' [*raconter des histoires*, as Derrida interprets the German *über Seiendes erzählende zu berichten* (the English edition has 'to report narratively'), which is translated literally, between parentheses, by '*informer en racontant*' (26)]" (26). Furthermore, he adds that narrative—namely, telling stories—has a specific meaning for Heidegger here: it accounts for "philosophy itself" as the ontic dissimulation of the question of Being and thus as "metaphysics and onto-theology" (26). This suggests that, despite the discrimination between Plato's and Aristotle's analysis of Being and Thucydides's narrative, the former are still on this side of philosophy as telling stories.[3]

To explain what telling stories means, Derrida alludes to a distinction between origin and genesis, which, as we will see, is at work in a key moment of the *Timaeus* and, on my reading, grounds the interpretation of Platonism elaborated in "Plato's Pharmacy." Derrida (2016) observes:

> To tell stories is . . . to assimilate being [*être*] and beings [*étant*], that is, to determine the origin of beings qua beings on the basis of another being. It is to reply to the question "what is the being of beings?" by appealing to another being supposed to be its cause or origin. It is to close the opening and to suppress the question of the *meaning* of being. Which does not mean that every ontic explication in itself comes down to *telling stories*; when the sciences determine causalities, legalities that order the relations between beings, when theology explains the totality of beings on the basis of creation or the ordering brought about by a supreme being, they are not necessarily telling stories. They "tell stories" when they want to pass their discourse off as the reply to the question of the meaning of being or when, incidentally, they refuse this question all seriousness. (29)

I highlight what interests us here: on the one hand, origin as the Being of beings, on the other, genesis as the transition from a being to another, as the becoming of things. Therefore, telling stories consists in the ontic explanation of the origin of beings. The recourse to the expression "telling stories," in order to interpret Heidegger's remark, is made explicit a little later on, when Derrida turns to section 2 in the "Introduction" of *Being and Time*. In the passage recalled by Derrida, Heidegger borrows from Plato's *Sophist* the determination of the ontic explanation of the origin of beings as a narrative, as telling a story.

> The being of beings "is" itself not a being. The first philosophical step in understanding the problem of being consists in avoiding the *mython tina diēgeisthai*, in not "telling a story," that is, not determining beings as beings by tracing them back in their origins to another being—as if being had the character of a possible being. (Heidegger 1996, 5)[4]

As Derrida observes, philosophy demarcates itself from "telling stories" when the Stranger in Plato's *Sophist* claims to abandon the mythological discourse in order to address the problem of Being as such.[5] Furthermore, in a remark on the translation of Heidegger's passage, Derrida draws attention to the present tense "consists," observing that telling stories is "a gesture that always threatens the question of being, yesterday, now and tomorrow" (2016, 31). The reading of "Plato's Pharmacy" that I propose below interrogates the irreducibility of this threat. Unfolding Heidegger's reference to the *Sophist*, in the seminar, Derrida explores how Plato takes the first philosophical step beyond mythology onto the question of Being. The renunciation of mythology is inscribed in the dialogue at the moment when, after the well-known refutation and parricide of Parmenides, the character of the Stranger sketches out a short history of past ontologies. Plato's text reads:

> As if we had been children, to whom they repeated each his own mythus or story [in the French edition quoted by Derrida: "*ils m'ont l'air de nous conter les mythes (muthon tina ekastos phainetai moi diēgeisthai)*," Derrida 2016, 32]; one said that there were three principles, and that at one time there was war between certain of them; and then again there was peace, and they were married and begat children, and brought them up; and another spoke of two principles, a moist and a dry, or a hot and a cold, and made them marry and cohabit. The Eleatics, however, in our part of the world, say that things are many in name, but in nature one; this

is their mythus, which goes back to Xenophanes, and is even older. Then there are Ionian, and in more recent times Sicilian muses, who have arrived at the conclusion that to unite the two principles is safer. (242c–e)

It is worth reading what Derrida adds at the end of the first case of ontology recalled by Plato. He suggests that what the latter tells is the history/story of being as "the history of being as a family history, as a family tree" (Derrida 2016, 32), thus alluding to the ideas of genesis and becoming. The conclusion of the Stranger's argument, Derrida summarizes, is that "Being is *other* than the determination of the *onta*" and thus "one must be *conscious* of this alterity which is not a difference between *onta*, in order to transgress mythology when one asks what is the origin of beings in their being" (34). However, Plato too admits that the task of abandoning mythology is impossible for the philosopher. Derrida evokes the example of Timaeus's preliminary remark in his discourse about the origin of the universe—about "the origin of the *world*, the origin of the beings" (35), as Derrida puts it—in *Timaeus* 27d–29d. This remark is interpreted by Derrida as a "response" to Socrates's demand for "a true story (*alēthinon logon*)" and not "a *muthon*" (35), which precedes the discourse. Approving his interlocutor's claim for the historical authenticity of the forthcoming discourse, Socrates observes: "The fact that it isn't a made-up story but a true historical account is of course critically important" (*Timaeus* 26e).[6] Timaeus begins by explaining that his discourse is marked by two related impossibilities: (a) the task of speaking about the father of the universe to everyone is impossible (28c); and, consequently, (b) it is impossible to give an account of the origin of the universe, namely, of the becoming of beings, that would be "altogether internally consistent and in every respect and perfectly precise" (29c), for the very reason that it regards becoming and not Being.[7] Derrida interprets Timaeus's remark at different levels: as a direct response to Socrates's observation, as a declaration of the impossibility of the ontological explanation of the origin of beings, and, finally, as a response to Heidegger's question about the language of destruction. Timaeus announces that "when it is a question of the origin of beings, a philosophical discourse adequate to the question is impossible," and "one must be content to recite [*réciter*], to unroll like [*dérouler comme*] a genesis, like a becoming-real of things, something that is not becoming, but the origin of things" (Derrida

2016, 35). Ultimately, "one must unroll the *Archē* like a genesis" (35). Here Derrida takes up the distinction between ontological origin and ontic genesis that he had referred to earlier, when designating the activity of "telling stories" as an ontic explanation of the ontological origin of things. I suggest that the emphasized expression of the ontic metaphor consists precisely in the *archē*'s irreducible developing *like* a genesis. Therefore, Timaeus's remark is interpreted "as the principle of an ironic answer to the question of being—in Heidegger's sense" (35). It accounts for the inescapable necessity of telling stories, of mythology, of the ontic metaphor.[8] Does this necessity also imply the impossibility of finding a grammar for the ontological task, and thus the irreducible relation between grammar as a concatenation of concepts and genesis as a concatenation of beings? I leave this double question open for the moment: the pages that follow may be read as an elaboration of it.

THE ORIGIN AND POWER OF THE LOGOS

The following interpretation of "Plato's Pharmacy" begins with an analysis of the opening scene of chapter 2, entitled "The Father of the Logos," where Derrida recalls the myth on which Socrates bases his indictment against writing in Plato's *Phaedrus*.[9] The myth is preceded by the formulation that "the story begins like this [*L'histoire commence ainsi*]" (Derrida 1981, 75), through which Derrida seems to suggest that the subsequent myth unfolds the story (/history) that the logos tells about its origin, about its originary and nonmetaphorical relation to its father. The story told by Socrates describes the scene in which the god Theuth presents the invention of the characters of writing (*grammata*) to the king of Egypt, Thamus.

> Theuth came to him and exhibited his arts and declared that they ought to be imparted to the other Egyptians. And Thamus questioned him about the usefulness of each one; and as Theuth enumerated, the King blamed or praised what he thought were the good or bad points in the explanation. Now Thamus is said to have had a good deal to remark on both sides of the question about every single art (it would take too long to repeat it here); but when it came to writing, Theuth said, "This discipline, my King, will make the

Egyptians wiser and will improve their memories: my invention is a recipe [*pharmakon*] for both memory and wisdom." But the King said . . . etc. (*Phaedrus* 274c–e)

Uncovering what is implicit and presupposed in this scene, Derrida observes that the story tells us about the origin of the logos—namely, the king, who neither knows about writing nor needs it, since he has the power of speech and is in the position of deciding about the value and utility of Theuth's invention. This position will be identified, in a moment, with that of the father. Derrida writes:

> The value of writing will not be itself, writing will have no value, unless and to the extent that god-the-king approves of it. But god-the-king nonetheless experiences the *pharmakon* as a product, an *ergon*, which is not his own, which comes to him from outside but also from below, and which awaits his condescending judgment in order to be consecrated in its being and value. God the king does not know how to write, but that ignorance or incapacity only testifies to his sovereign independence. He has no need to write. He speaks, he says, he dictates, and his word suffices. Whether a scribe from his secretarial staff then adds the supplement of a transcription or not, that consignment is always in essence secondary. (1981, 76)

Therefore, the *parti pris* of this scene is the relationship between the "origin and power of speech, precisely of *logos*" and the "paternal position" (76)—that is, the story about the origin of the logos as the position of the king-father. Derrida summarizes this story through the following formulation: "The origin of logos is *its father*" (77). This means that the position of the father is understood as the subject's power to speak and thus to emit and accompany a logos. From this perspective, the father is not a metaphor insofar as he does not result from the importation of the genetic relation from the zoological discourse to the linguistic one. The father and the logos are as such by virtue of their originary and nonmetaphorical relationship. To this extent, the logos is a son as it is emitted and accompanied by its father. This thesis is anything but trivial as it accounts for the structure of the logos in general, from linguistics to zoology. Furthermore, it is precisely from this perspective that, by definition, writing entails the disappearance of the father, whether natural or violent.[10] Derrida observes:

> Not that logos *is* the father, either. But the origin of logos is *its father*. One could say anachronously that the "speaking subject" is the *father* of his speech. And one would quickly realize that this is no metaphor, at least not in the sense of any common, conventional effect of rhetoric. *Logos* is a son, then, a son that would be destroyed in his very *presence* without the present *attendance* of his father. His father who answers. His father who speaks for him and answers for him. Without his father, he would be nothing but, in fact, writing. At least that is what is said by the one who says: it is the father's thesis. (1981, 77)

Why must we say the origin of the logos by referring to the genitor, to the supposedly zoological metaphor of generation? What does this necessity mean? Indeed, these questions have already been eluded since the relationship between the logos and the father is understood not as a metaphor but as the structure of the logos in general. Derrida illustrates this structure by drawing attention to the determinations that are attributed to the *logoi* in the *Phaedrus*. They are designated as noble creatures—namely, as creatures of noble birth or race (*gennaioi*), and as sons of the subject that pronounces and protects them (*patēr*).[11]

The logos, this noble creature, does not only belong to a system of discourses; rather, it is the very element of the system itself. Holding on to Socrates's description of the logos as a living being (*zōon*) in *Phaedrus* 264b–c,[12] Derrida suggests that the logos-*zōon* is the object of linguistics as well as of zoology. In other words, it is the minimal particle, the atom, of these regional discourses. "*Logos* is a *zōon*," Derrida observes, "an animal that is born, grows, belongs to the *physis*. Linguistics, logic, dialectics, and zoology are all in the same camp [*ont partie liée*]" (1981, 79). As remarked by Socrates in his description, the logos-*zōon* has a body proper and is not deformed. It is "an *organism*," Derrida explains, "a differentiated body *proper*, with a center and extremities, joints, a head, and feet" (80). This suggests once more that the structure of the logos constitutes a metaphor borrowed from a certain understanding of the living and thus that the relation to its father (the noble birth, the body proper, etc.) hinges on a genetic and zoological explanation. However, here Derrida relaunches the question of the father. He explains that the relation between the logos and its father does not consist in the metaphorical inscription of the zoological relationship between the son and its genitor into the linguistic discourse. Rather, it is the element of the metaphorical exchange between the regional discourses of the system. The father is not the genetic cause of the living son, nor does the birth of the logos constitute a generation, since the relation between the father and the logos is not metaphorical but originary. In other words, it is this relation that makes them what they are, logos and father. Therefore, the living being is understood on the basis of the logos, and its generation consists in the relation to the power of the logos. Derrida argues:

> One would then say that the origin or cause of *logos* is being compared to what we know to be the cause of a living son, his father. One would understand or imagine the birth and development of *logos* from the standpoint of a domain foreign to it, the transmission of life or the generative relation. But the father is not the generator or procreator in any "real" sense prior to or outside all relation to language. In what way, indeed, is the father/son relation distinguishable from a mere cause/effect or generator/engendered relation, if not by the instance of logos? Only a power of speech can have a father. The father is always father to a speaking/living being. In other words, it is precisely *logos* that enables us to perceive and investigate something like paternity. (80)

According to a logic that seems to invert the appearances, what is the most familiar—the very concept of family—is grounded on the originary and nonmetaphorical relation between the logos and its father and not on the genetic and zoological relation between the son and its genitor. Proposing a formulation whose implications extend throughout his early work, Derrida remarks that the concept of family rests on a linguistic rather than zoological element.[13] "If there were a simple metaphor in the expression 'father of logos,'" Derrida observes, "the first word, which seemed the more *familiar*, would nevertheless receive more meaning *from* the second than it would transmit *to* it. The first familiarity is always involved in a relation of cohabitation with *logos*" (81). Going back to Socrates's description of the logos-*zōon*, we may conclude that the logos-*zōon* does not depend on a zoological metaphor but on the presupposition of the paternal thesis. Therefore, the speculation on the question of the father ends up calling for a reconsideration of the metaphorical organization of the system:

> To have simple metaphoricity, one would have to make the statement that some living creature incapable of language, if anyone still wished to believe in such a thing, has a father. One must thus proceed to undertake a general reversal of all metaphorical directions, no longer asking whether *logos* can have a father but understanding that what the father claims to be the father of cannot go without the essential possibility of *logos*. (81)

The system described here is what Derrida designates elsewhere as the *logos spermatikos*, in which the concepts of the living and of the zoological process of generation are grounded on the concept of logos and on the relationship between the logos and its subject, respectively.[14] The immediate implication of this system is that a living being is what it is only if it bears within itself the power of the logos and thus it is accompanied by its father.[15]

THE TEXTUALITY OF PLATO'S TEXT

In chapter 3, entitled "The Inscription of the Sons," Derrida draws attention to "the structural resemblance between the Platonic and the other mythological figures of the origin of writing" (1981, 86). The consequence of this operation, as he points out, consists in casting light on the relation between the myth and the logos and on the myth that the logos tells about its origin as the position of the father.[16] In particular, I refer to the pages in which Derrida explains that the Egyptian Thot is a "spokesman" (88) of Ra, "the god of creative word," and replaces it "only by metonymic substitution, by historical displacement, and sometimes by violent subversion" (89).[17] Derrida suggests that the substitution of Ra with Thot, which occurs within the element of linguistic permutations or concatenations (within the limits of grammar, we may say), cannot be understood as merely an inoffensive word play since Thot is often involved in "plots, perfidious intrigues, conspiracies to usurp the throne." As Derrida explains, "He helps the sons do away with the father, the brothers do away with the brother that has become king" (90). Does this mean that there is no such a thing as the father, origin and power of the logos, and thus that *logoi* without a father have already replaced and usurped one another? In other words, does this mean that there is anything but genealogical breaks? Indeed, at a certain point, Thot becomes the god of creative speech, Ra. "The same can also be seen to occur in the evolution [*comme une évolution dans* . . .] of the history of mythology" (91), Derrida remarks in a note. Therefore, the logos itself, understood as Socrates's logos-*zōon*, as the element of the system called Platonism, consists in a genealogical break, a usurpation, that tells the story/history of its originary relation to the father.

In chapter 4, entitled "The Pharmakon," Derrida remarks that in the word *pharmakon* are tied together the threads of the correspondence between the Egyptian Thot and the character of Socrates's myth.[18] "My invention is a recipe (*pharmakon*) for both memory and wisdom," Theuth says according to Socrates. Derrida observes that "the [French] translation [of the word *pharmakon*] by 'remedy' [*remède*] erases [*efface*], in going outside the Greek language, the other pole reserved in the word *pharmakon*" (97).[19] The operation of erasing (*effacer*), Derrida explains, consists in the annihilation of the ambiguity—that is, of the uninterrupted communication, between the opposite meanings of *remedy* and *poison*, both of which are gathered together in the unity of the same signifier, *pharmakon*.[20] But what, precisely, does this ambiguity and communication account for and, consequently, what does the translation by "remedy" destroy? The textuality of Plato's text, Derrida answers—namely, the text's anagrammatic structure, which carries the multiple meanings of a word and the uninterrupted ambiguity and communication between them.

> The effect of such a translation is most importantly to destroy what we will later call Plato's anagrammatic writing, to destroy it by interrupting the relations interwoven among different functions of the same word in different places, relations that are virtually but necessarily "citational." When a word inscribes itself as the citation of another sense of the same word, when the textual center-stage of the word *pharmakon*, even while it means *remedy*, cites, re-cites, and makes

> legible that which *in the same word* signifies, in another spot and on a different level of the stage, *poison* (for example, since that it is not the only other thing *pharmakon* means), the choice of only one of these renditions by the translator has as its first effect the neutralization of the citational play, of the "anagram," and, in the end, quite simply of *the very textuality of the translated text* [my emphasis]. (98)

As demonstrated by Derrida in *Of Grammatology*, the anagram consists in the structure of general writing that is necessarily presupposed by the phoneme, and thus in the irreducible synthesis of the grapheme.[21] From the perspective of the analysis developed in the previous section, anagrammatic structure contests the dissociation of writing and speech as well as the paternal position of the logos-*zōon* as the element of the system of discourses and metaphorical exchanges called Platonism. Furthermore, anagrammatic structure intersects what Hegel identifies as the speculative resources of the German language, which sometimes employs the same word for opposed significations. Later, I develop this reference, which can be tracked in key moments across Derrida's early work.[22] Focusing on the text we are reading, I note that anagrammatic structure holds in reserve the different functions—namely, the multiple meanings that a grapheme takes on according to the multiple concatenations in which it is reinscribed. In other words, anagrammatic structure ties together the grammatical or syntactical concatenations that precede the determination of the meaning of a word (for instance, of *pharmakon*) and thus the destruction of its structural ambiguity.[23] As suggested by the verb *ré-citer*, which Derrida uses to account for the relation of the word to its meanings, these grammatical concatenations may be interpreted as stories, as discourses that are repeated and thus are not accompanied by their father. Therefore, the irreducible synthesis of grammatical concatenations and stories that make up the grapheme *pharmakon* constitutes the very element of Platonism, the vigil from which it wishes to dissociate itself. Derrida makes this explicit when he contends that the destruction of the ambiguous and anagrammatic writing in the translation by "remedy" is already "an effect of Platonism," "the consequence of something already at work . . . in the relation between Plato and his language" (98). The fact that a text aims to destroy its textuality, its anagrammatic structure, cannot be excluded. Rather, it is a work that textuality makes possible and, at the same time, it is never accomplished so long as the communication between the grammatical concatenations that constitute the anagrammatic structure of a text can never be fully destroyed. "Textuality being constituted by differences and by differences from differences," Derrida writes, "it is by nature absolutely heterogeneous and is constantly composing with the forces that tend to annihilate it" (98). It is time to recall Thamus's response to Theuth:

> Most artful Theuth, one person is able to bring forth the things of art, another to judge what allotment of harm and of benefit they have for those who are going to use them. And now you, being the father of written letters, have on account of goodwill said the opposite of what they can do. For this will provide forgetfulness in the souls of those who have learned it, through neglect of memory, seeing that, through trust in writing, they recollect from outside with alien markings, not reminding themselves from inside, by themselves. You have therefore found a drug not for memory, but for reminding. You are supplying the opinion of wisdom to the students, not truth. (*Phaedrus* 274e–275b)

As Derrida explains, Plato, through Thamus, wishes to master the ambiguity of the *pharmakon* by establishing a system of rigid oppositions (good and evil, inside and outside, true and false, essence and appearance). Within this system, he remarks that "writing is essentially bad, external to memory, productive not of science but of belief, not of truth but of appearances" (1981, 108). The line between the opposites drawn by Thamus also demarcates memory (*mnēmē*) from re-memoration (*hypomnēsis*)—that is, as Derrida designates them, "an unveiling (re-)producing a presence" from "the mere repetition of a monument" ("archive"), "truth"/"being" and "sign"/"type" (108–9), etc. At least, this is Plato's dream: a *mnēmē* dissociated from *hypomnēsis*. Derrida argues against this dissociation, showing that the minimal structure of memory as well as of the living organism is a written sign or an inscription. The latter allows the living being, which is finite—namely, the living organism—to relate itself to the nonpresent, and thus it puts memory to work.

> Memory is finite by nature. Plato recognizes this in attributing life to it. As in the case of all living organisms, he

assigns it, as we have seen, certain limits. A limitless memory would in any event be not memory but infinite self-presence. Memory always therefore already needs signs in order to recall the nonpresent, with which it is necessarily in relation. The movement of dialectics bears witness to this. Memory is thus contaminated by its first substitute: *hypomnēsis*. (109)

This reading is based on the definition of the living organism that Plato has in *Timaeus* 89c. A few pages prior, Derrida recalls this definition as follows: "It [the living] has a limited lifetime . . . death is already inscribed and prescribed within its structure, in its 'constitutive triangles'" (101).[24] Furthermore, he measures the limited lifetime of the living organism against the "immortality and perfection," which, according to *Republic* 2.381b–c, "would consist in its having no relation at all with any outside" and is accomplished only in God (1981, 101). Therefore, the inscription in the memory is the minimal structure of the living organism to the extent that the latter is finite and relates to the non-present only through memory.

AUTOCHTHONY

In the chapter examined above, Derrida parses that the system of opposition evoked by Thamus's response to Theuth presupposes the very concept of opposition, the matrix of all oppositions, which consists in the line drawn between the inside and the outside and dividing them. He explains:

> It is not enough to say that writing is conceived out of this or that series of oppositions. Plato thinks of writing, and tries to comprehend it, to dominate it, on the basis of *opposition* as such. In order for these contrary values (good/evil, true/false, essence/appearance, inside/outside, etc.) to be in opposition, each of the terms must be simply external to the other, which means that one of these oppositions (the opposition between inside and outside) must already be accredited as the matrix of all possible opposition. And one of the elements of the system (or of the series) must

also stand as the very possibility of systematicity or seriality in general. (103)

This insight about the foundational opposition between the inside and the outside is unfolded in chapter 6, entitled "The Pharmakos." The opening scene of the chapter recounts the myth that the logos tells about its origin (father, noble birth, body proper) and through which it wishes to destroy the other myths that it bears within itself—namely, its own textuality—and to demarcate itself from them. Here the opposition between the inside and the outside is described as the very institution of the myth of the logos and thus of logic itself.

> The purity of the inside can then only be restored if the *charges are brought home* against exteriority as a supplement, inessential yet harmful to the essence, a surplus that *ought* never to have come to be added to the untouched plenitude of the inside. The restoration of internal purity must thus reconstitute, *recite*—and this is myth as such, the *mythology* for example of a *logos* recounting its origin, going back to the eve of the pharmakographic aggression—that to which the *pharmakon* should not have had to be added and attached like a *literal parasite*: a *letter* installing itself inside a living organism to rob it of its *nourishment* and to *distort* the pure audibility of a voice. Such are the relations between the writing supplement and the *logos-zōon*. In order to cure the latter of the *pharmakon* and rid it of the parasite, it is thus necessary to put the outside back in its place. To keep the outside out. *This is the inaugural gesture of "logic" itself* [my emphasis]. (128)

In the subsequent analysis, Derrida suggests that this inaugural gesture is political as it accounts for the constitution of the political community of the city. He highlights the systematic link between the word *pharmakos*, which is not used by Plato, and the lexicon of the *pharmakon*. *Pharmakos* is a synonym of *pharmakeus*—namely, the magician or the one who gives the *pharmakon*—a title that Plato attributes to Socrates, as pointed out in the previous chapter of "Plato's Pharmacy."[25] However, the word has a peculiarity that captures Derrida's interest: it is "the unique feature of having been overdetermined, overlaid by

Greek culture with another function . . . another role, and a formidable one" (130). Relying on the historiographical sources available at the time of the publication of "Plato's Pharmacy," Derrida points out that the *pharmakos* is involved in the ritual practices of purification of a community and its major significations are "the *evil* and the *outside*, the expulsion of the evil, its exclusion out of the body (and out) of the city" (130).[26] In Athens, it enters the stage during the celebrations of the public ceremonies of the Thargelias, which comprised the rites of purification of the city. These rites included the sacrifice of two deformed individuals as a remedy for the calamity that affected the city.[27]

Derrida recounts the story transmitted by his sources by articulating the ritual of the *pharmakos* with the myth told by Socrates in the *Phaedrus* and, more generally, with the myth about the origin of the logos, the paternal thesis of Platonism. In the ritual, what is designated by the "city" or the "community" of the Athenians is the institution of the inside and thus the description of the line that separates the pole of the logos-*zōon* (noble birth, body proper, and so forth) from the opposite pole of the *pharmakoi*. Derrida writes:

> The city's body *proper* thus reconstitutes its unity, closes around the security of its inner courts, gives back to itself the word that links it with itself within the confines of the agora, by violently excluding from its territory the representative of an external threat or aggression. That representative represents the otherness of the evil that comes to affect or infect the inside by unpredictably breaking into it. Yet the representative of the outside is nonetheless *constituted*, regularly granted its place by the community, chosen, kept, fed, etc., in the very heart of the inside. These parasites were as a matter of course domesticated by the living organism that housed them at its expense. (133)[28]

The political gesture that is at the origins of logic and of the system of the logos-*zōon* can be understood as "autochthony." This word occurs only once in the "Pharmacy," as a determination that opposes the Socratic logos to the errancy of writing. "The Socratic word," Derrida writes, "does not wander, stays at home, is closely watched: within *autochthony* [my emphasis], within the city, within the law, under the surveillance of its mother tongue" (124). A deeper examination of the meanings of this word is developed by Derrida in later writings, which also rely on Nicole Loraux's studies on the origins of Athens. The next chapter explores the relationship between autochthony and *khōra* as Derrida elaborates it in the reading of Plato's *Timaeus* that he undertakes throughout his work.[29]

THE NATURAL TENDENCY TO DISSEMINATION

Derrida takes up the interrupted thread of the reading of the *Phaedrus* in chapter 8, entitled "The Heritage of the Pharmakon. Family Scene." As is well known, Socrates compares the written *logoi* to the offspring of painting and remarks that, when interrogated about what they say, they remain silent and limit themselves to indicating the same (*Phaedrus* 275d). From this, he concludes that they cannot protect themselves but demand the assistance of the father (275e). Derrida points out that Socrates speaks about writing from within the system of the logos-*zōon*, as a kind of logos that has been written down ("what is written down is a *logos* [*un* discours *écrit*]," Derrida 1981, 143) and thus dissociated from the presence of the father. Within this system, the written logos is not a noble creature but is "deformed at its very birth" (148) and thus constitutes a virtual *pharmakos*. It is measured against its brother of noble birth, the logos inscribed in the soul of "the one who understands," which has "the power to defend itself" (*Phaedrus* 276a)—namely, a father—and thus consists in the element of the system of regional discourses or *logoi*.[30] As Derrida remarks, Socrates refers to the graphic metaphor to account for the noble brother of the written logos and, therefore, for the supposedly nonmetaphorical element that presides over the metaphorical exchanges of the system. What renders this metaphor necessary, although this necessity is eluded by Plato here, is, for instance, that the written sign or grapheme links the living organism to the non-present and constitutes the elementary structure of memory and life. "This borrowing is rendered necessary," Derrida writes, "by that which structurally links the intelligible to its repetition in the copy, and the language describing dialectics cannot fail to call upon it" (1981, 149).[31] Anticipating what follows in Plato's text, Derrida proposes a distinction between the two *logoi*/brothers that allows him to bring to the fore, for the first time in his published work, the word "dissemination":

It is later confirmed that the conclusion of the *Phaedrus* is less a condemnation of writing in the name of present speech than a preference for one sort of writing over another, for the fertile trace over the sterile trace, for a seed that engenders because it is planted inside over a seed scattered wastefully outside: at the risk of *dissemination.* (149)

This is a metaphor insofar as the logos inscribed in the soul, the element of all metaphorical exchanges, grounds the zoological discourse about the living and the process of generation. From this perspective, the seed at risk of dissemination is understood in relation to writing as a written logos, as a logos deprived of the noble birth as well of the body proper, as a virtual *pharmakos*. In principle, only the logos-*zōon* is generative as it carries with itself the presence and assistance of the father and consists in an engendered organism. Therefore, generation should take place only within the limits of the community (family or city).[32]

Socrates suggests that writing is like sowing seeds through a pen, seeds that have no power to defend themselves nor to teach anything (*Phaedrus* 276b–c). In so doing, he establishes an analogy between two kinds of writing and two kinds of seeds, which Derrida describes as, respectively, "superfluous seeds giving rise to ephemeral produce (floriferous seeds)" and "strong, fertile seeds engendering necessary, lasting, nourishing produce (fructiferous seeds)" (Derrida 1981, 151). The consequence of the logos-seed analogy, which goes—against all appearances—from the logos to the seed, is that nature is constituted according to the system of the logos-*zōon*. Therefore, Derrida associates the aforementioned passage from the *Phaedrus* with a text from the *Laws* in which the character of the Athenian describes a law proposal against pederasty and the practices of sex that do not lead to fecundation. Here the implicit figure of the generative seed within the limits of the family seems to draw together law and nature. Plato's text, quoted by Derrida, reads:

That was exactly my own meaning when I said I knew of a device for establishing this law of restricting procreative intercourse to its natural function by abstention from congress with our own sex, with its deliberate murder of the race and its wasting of the seed of life on a stony and rocky soil, where it will never take root and bear its natural fruit, and equal abstention from any female field whence you would desire no harvest. (152–53)

Unfolding Derrida's reference, I argue that, from the perspective of the logos-*zōon* and thus of the generative seed of the family, dissemination is understood as contrary to nature as well as to law. However, there is also a place in Plato's corpus, once again in the *Timaeus*, where the germ of the deconstruction of the law-nature articulation and, more generally, of the whole system may be found. It is the description of sperm included in Timaeus's discourse, which Derrida recalls without developing its implications. He limits himself to recognizing that "the natural tendency of sperm is opposed to the law of *logos*" (154). The passage explains that the seed is constituted by a vital drive to go out of itself (and to generate) and, for this reason, the living resists the logos. If we link this drive to dissemination, Timaeus's explanation suggests that we can no longer think of the living in relation to the logos-*zōon* and to the latter's determinations. Timaeus observes:

> The marrow . . . we have named semen. And the semen, having life and becoming endowed with respiration, produces in that part in which it respires a lively desire of emission, and thus creates in us the love of procreation [*generation*]. Wherefore also in men the organ of generation becoming rebellious and masterful, like an animal disobedient to reason [*tou logou*], and maddened with the sting of lust, seeks to gain absolute sway. (*Timaeus* 91b)

What is at stake in this passage is an understanding of dissemination as the irreducible structure of the living, an understanding that calls into question the system designated as Platonism and based on the logos-*zōon* and the generative seed. Generation itself is made possible only by dissemination and thus before the inaugural gesture of logic, the institution of the city, the living-*zōon*, and so forth. If the logos has already been written down and writing has already been dissociated from its father—namely, disseminated—then the grapheme-seed is the element of linguistics, zoology, politics, and thus of all regional discourses.

Returning to the *Phaedrus*, Derrida recalls that Socrates also takes into account the case in which discourses are written at the service of dialectics and thus of the truth. In this case, Derrida suggests, they can be read as a "dialectical trace" opposed to a "non-dialectical trace" (Derrida 1981, 194). Here Plato's text seems to unfold the concepts of generation and life as they are at work within the system of the logos-*zōon* (namely within the horizon of the *logos spermatikos*). Socrates observes:

> When someone using the dialectical art, taking hold of a fitting soul, plants and sows with knowledge speeches that are competent to assist themselves and him who planted and are not barren but have seed, whence other speeches, naturally growing in other characters, are competent to pass this on, ever deathless, and make him who has it experience as much happiness as is possible for a human being. (*Phaedrus* 276d–277a)

Therefore, life can be transmitted and conserved only through the *logoi* inscribed by the dialectical man in fitting souls, *logoi* that are living, noble, with a body proper, and thus, in principle, fertile and generative.

THE SCIENCE OF THE DISSEMINATED TRACE

The hypothesis of the inscription-seed is put to the test in the reading of the passage from the *Timaeus* where *khōra* enters the stage. Derrida returns to the question of the origin of the world that he had addressed a few years earlier in the Heidegger course. In this case, he identifies the "origin of the world" with the "trace"—namely "the inscription of forms and schemes in the *matrix*, in the receptacle [this passage is partially missing in the English translation]" (1981, 159–60). The trace is disseminated, as Derrida suggests at the end of the long quotation from *Timaeus* 48e–51b, remarking that "the *khōra* is big [*grosse*] with everything that is disseminated here [*qui se dissémine ici*]" (161). Therefore, the disseminated trace is the origin of the world, or, as we know, the particle in which linguistics, politics, and cosmology are interwoven together. It is from this perspective that I propose to interpret the double determination that Derrida attributes to Timaeus's trace: "the *production of the son* and at the same time the constitution of *structurality*" (161). The disseminated trace constitutes the minimal structure of linguistic as well as zoological discourses: it has already been inscribed in the system of functions (concatenations, stories, senses, etc.) and relations among different functions, which is called textuality.

Derrida seems to suggest that we borrow from Plato himself the name of the science of the trace as the origin of the world. He recalls that Theuth also enters Plato's text, as the inventor of grammar, in the *Philebus*: as Derrida puts it, "of grammar, of the science of grammar as a science of differences" (162). In the passage from the *Philebus* recalled by Derrida, Plato identifies grammar as the science of letters and of their concatenations:

> The unlimited variety of sound was once discerned by some god, or perhaps some godlike man; you know the Story that there was some such person in Egypt called Theuth. He it was who originally discerned the existence, in that unlimited variety, of the vowels . . . and then of other things which, though they could not be called articulate sounds, yet were noises of a kind. There were a number of them, too, not just one, and as a third class he discriminated what we now call the mutes. Having done that, he divided up the noiseless ones or mutes until he got each one by itself, and did the same thing with the vowels and the intermediate sounds; in the end he found a number of the things, and affixed to the whole collection, as to each single member of it, the name "letters." It was because he realized that none of us could get to know one of the collection all by itself, in isolation from all the rest, that he conceived of "letter" as a kind of bond of unity uniting as it were all these sounds into one, and so he gave utterance to the expression "art of letters," implying that there was one art that dealt with the sounds. (*Philebus* 18b–d)

This science, the art of grammatical concatenations, provides the framework of the science of textuality, of the anagrammatic structure in which Plato's text is inscribed and from which it wishes to demarcate itself as the system of the logos-*zōon*, or the *logos spermatikos*. Here we discover the degree to which Derrida developed the opening question of *Being and Time* about the syntax of ontology. This understanding of grammar is unfolded in an explicit fashion through the reading of the *Sophist* that Derrida proposes in the last pages of "Plato's Pharmacy." He returns again, and not by chance, I suggest, to the examined session of the earlier lecture course where Plato dissociates philosophy from mythology, the latter of which is identified with the history of Greek thought. In particular, Derrida draws attention to Plato's recourse to the grammatical science as an explanatory concept, at a precise moment of the discourse of the Stranger.

> Is it then by chance if, once "being" has appeared as a *triton ti*, a third irreducible to the dualisms of classical ontology, it is again necessary to turn to the example of grammatical science and of the relations among letters in order to explain the interlacing that weaves together the system of differences (solidarity-exclusion), of kinds and forms, the *sumplokē tōn eidōn* to which "any discourse we can have owes its existence" (*ho logos gegonen hēmin*) [*le discours nous est né*] (259e)? (165)

The Stranger explains the dynamic of the concatenation among genres and forms by referring to the example of the concatenation among the letters. "This communion of some with some," he observes, "may be illustrated by the case of letters; for some letters do not fit each other, while others do" (*Sophist* 253a).[33] This dynamic is not immediately grasped by everybody, but is the object of a science, the grammatical science (253a). Now, if the five fundamental genres undergo an analogous dynamic of concatenation, the Stranger continues, this dynamic should be, in turn, the object of a science, the dialectical science, which would be the greatest of all sciences (253b–d).[34] Therefore, the Stranger conceives of dialectics in the light of grammar, at the same time as he subordinates the latter to the former. As Derrida remarks, what ultimately justifies, for Plato, the primacy of dialectics over grammar is the paternal thesis and thus the concept of the logos-*zōon*. Dialectics is dissociated from grammar, he summarizes, as it is "guided by an intention of *truth*" and can be fulfilled only "where truth is fully present and fills the *logos*" (Derrida 1981, 166). Therefore, by affirming the primacy of dialectics over grammar, the work of Plato against the textuality (that is, the anagrammatic structure) of his text has already begun. Indeed, Derrida points out that what the Stranger rejects with the parricide of Parmenides is not only the impossibility of the full and absolute presence of what is, of a full intuition of the truth, but precisely the fact "that the very condition of discourse—*true or false*—is the diacritical principle of the *sumplokē*" (166).[35]

Ultimately, it is not by chance that, as Derrida implicitly suggests in the aforementioned passage, Plato's definition of the *symplokē* may be read from the point of view of dialectics—namely, of the logos-*zōon* and the system grounded on it—as well as from the point of view of grammar: that is, of dissemination and the trace-seed. Here we may find an account of the genesis—of the origin of the world as a disseminated trace—that is presupposed by linguistics, zoology, politics, and so forth. Plato's text reads:

> The attempt at universal separation is the final annihilation of all reasoning; for only by the union of conceptions with one another do we attain to discourse of reason [*o logos gegonen ēmin*, quoted by Derrida as "*le discours nous est né*"]. (*Sophist* 259e)

2

PLATONISM II
Khōra

> *Khōra* would make or give *place*; it would give rise—without ever giving anything—to what is called the coming of the event. *Khōra* receives rather than gives. Plato in fact presents it as a "receptacle." Even if it comes "before everything," it does not exist for itself. Without belonging to that to which it gives way or for which it makes place [*fait place*], without *being a part* [*faire partie*] of it, without *being of it*, and without being something else or someone other, giving nothing other, it would give rise or allow to take place.
>
> —Derrida, *Rogues: Two Essays on Reason*, xiv

In chapter 6 of "Plato's Pharmacy," Derrida describes Platonism as the philosophical system of autochthony called Athens, in which the logos-*zōon* and its originary and nonmetaphorical relation to its father constitute the element of all discourses, from zoology to politics. However, there is a moment in Plato's text, Derrida suggests, once more in the *Timaeus*, where a beyond of Platonism is conjured up, demanded by a necessity that remains to be explained. It is the moment in which Timaeus acknowledges that he must go back and recommence his discourse about the origin of the universe, this time by taking into account a third element beside the intelligible paradigm and the latter's sensible imitation—namely, *khōra* (*Timaeus* 48a–52d).

Derrida puts forth a preliminary reading of the recourse to *khōra* in the *Timaeus* in chapter 9 in "Plato's Pharmacy." As we know, he summarizes the necessity of this recourse as follows: "All these things [that Timaeus had exposed in his cosmological discourse] 'require' (*Timaeus* 49a) that we define the origin of the world as a *trace*, that is, the inscription of forms and schemes in the matrix, in the receptacle [this passage is partially missing in the English translation]" (1981, 159–60). Developing this necessity, Derrida continues that "it is a matrix, womb, or receptacle that is never and nowhere offered up in the form of presence, or in the presence of form, since both of these already presuppose an inscription within the mother" (160). A quotation from the French translator of Plato's text (Rivaud) follows: it highlights the impossibility of accounting for what Plato designates by "place" without resorting to further metaphors, and, simultaneously, the confusion that this designation had produced in modern readers. Finally, Derrida formulates a hypothesis about *Timaeus* 48e–51b that would mark his work thoroughly: "Here is the passage beyond all 'Platonic' oppositions, toward the aporia of the originary inscription" (160). From the perspective of "Plato's Pharmacy," we know that the evoked passage refers to the textuality of the Platonic text, to the latter's anagrammatic structure and to the dissemination of the trace-seed as the element of discourse as well as of the living. In this chapter, I examine how Derrida develops his hypothesis in later texts that are explicitly dedicated to the interpretation of the aforementioned passage from the *Timaeus*. I refer to the unedited 1970–1971 lecture course on *Théorie du discours philosophique: Conditions de l'inscription du texte de philosophie politique—l'exemple du matérialisme (Theory of Philosophical Discourse: Conditions for the Inscription of the Text of Political Philosophy— The Example of Materialism)*; to the unedited 1985–1986 seminar on *Littérature et philosophie comparées: Nationalité et nationalisme philosophique: Mythos, logos, topos (Comparative Literature and Philosophy: Nationality and Philosophical Nationalism)*; and to the essay entitled *Khōra* (first published in 1987, in a collection in homage to Jean-Pierre Vernant and then revised and published separately in 1993). Ultimately, I demonstrate that the interrogation of *khōra* throughout these texts allows Derrida to think the general condition for generation and history and, thus, the very concept of history.

This analysis of *khōra* across time consists of a heterogeneous process of stratification in which different texts echo and interpolate one another. The 1985–1986 texts dedicated to *khōra* represent the most problematic source that is being analyzed here, as they are transcriptions of a series of scattered lectures taught by Derrida. From these texts, I focus on those passages that are anticipated in the earlier remarks and/or return in the later *Khōra*. As we will see, Derrida's recursive approach to *khōra* is prescribed and programmed by the very discourse of Timaeus on *khōra* (more precisely, by *khōra* itself, in which this discourse too is reflected) and by the structural law that Derrida's reading extracts from this discourse. Within the framework of this law, Derrida marks the singularity of his own reading.[1]

THE EARTH OF FATHERS

Before delving in the reading of Derrida's analysis of the *Timaeus*, I highlight the function that *khōra* plays in Athens, as Nicole Loraux analyzes it in *The Children of Athena* (1981). This analysis, which follows Derrida's first texts on *khōra*, is evoked by Derrida himself in his subsequent works, also to support his original interpretation of the *Timaeus* and, more generally, of Platonism.

The earliest references to Loraux's work and, in particular, to chapter 1 ("Autochthony: An Athenian Topology") and to an important passage from chapter 3 ("The Athenian Name"), appear in two texts from the 1985–1986 lectures dedicated to *khōra*. In the first text, entitled "Questions. Le 11 décembre 1985," Derrida refers to *The Children of Athena* by summarizing Loraux's reconstruction of the double function of the civic soil—namely, *khōra*. The Athenians were born from the soil of their fatherland (autochthony) and return there when they die, through inhumation.[2] Derrida quotes a passage from Loraux's text in which this double function is seen as accomplished in two different places of the civic soil of Athens, the Acropolis and the Kerameikos:

> On the Acropolis, a *son of the earth* is born, the king whom "Athena, daughter of Zeus, brought up in an earlier age and established in her own rich sanctuary"; this Erechtheus, the "child of the fertile loam," whose cult is recalled in the *Iliad*, this Erichthonius whom, on Athenian vases, Ge brings forth into the light, not far from the symbolic olive tree. At each celebration of the Panathenea, the history of Athens begins,

or begins anew. In the Kerameikos are buried the *sons of the fatherland* in the civic soil (*chōra*) from which they have sprung, men for whom time is erased through an endless return, at their ends, to their origins. Thus, from the sacred hill, to the official cemetery, a gap opens between two "discourses" on the subject of autochthony, one consisting of ritual and figural representations, the other the secularized discourse of political prose. As if each site produced its own language. As if autochthony changed its form with each shift in the civic space where it is expressed. (Loraux 1993, 42)[3]

Furthermore, Derrida links this passage to the text from Plato's *Menexenus* (237b) in which, through a discourse attributed to the courtesan Aspasia, Socrates explains the noble birth (*eugeneia*) of the Athenians with the fact that they are literally born from *khōra*. Although Plato's text is not quoted in Loraux's passage, it constitutes one of the key sources of the analysis of autochthony developed throughout her book. Plato writes:

> And first as to their birth. Their ancestors were not strangers, nor are these their descendants sojourners only, whose fathers have come from another country; but they are the children of the soil [*khōra*], dwelling and living in their own land. And the country [*patris*] which brought them up is not like other countries, a stepmother to her children, but their own true mother; she bore them and nourished them and received them, and in her bosom they now repose. It is meet and right, therefore, that we should begin by praising the land which is their mother, and that will be a way of praising their noble birth. (*Menexenus* 237b)

This passage makes explicit why Derrida is interested in Loraux's reading of the *Menexenus*. Autochthony seems to account for the system in which the citizens of Athens entertain an originary relationship with their fatherland (*khōra/patris*), a relationship that is not a metaphor borrowed from the biological process of generation—rather, the opposite holds for Plato, as we will see—and grants them their *eugeneia*. The second reference appears in the pages of the text entitled "CHORA (suite)," in which Derrida interweaves his reading of Kantorowicz's

"Mysteries of State: An Absolutist Concept and its Late Medieval Origins" (1955) and Loraux's remarks on Plato's *Menexenus*, especially on the tradition of the funeral oration that Socrates takes up in his praise for the Athenian soldiers who died in war. After paraphrasing Socrates's first praise for the fatherland, Derrida recalls a second praise that, according to Socrates, should be addressed to the fatherland "by all mankind" (*Menexenus* 237c):[4]

> The second praise which may be fairly claimed by her is that at the time when the whole earth was sending forth and creating diverse animals, tame and wild, she our mother was free and pure from savage monsters, and out of all animals selected and brought forth man, who is superior to the rest in understanding, and alone has justice and religion. And a great proof that she brought forth the common ancestors of us and of the departed, is that she provided the means of support for her offspring. For as a woman proves her motherhood by giving milk to her young ones (and she who has no fountain of milk is not a mother), so did this our land prove that she was the mother of men, for in those days she alone and first of all brought forth wheat and barley for human food, which is the best and noblest sustenance for man, whom she regarded as her true offspring. And these are truer proofs of motherhood in a country than in a woman, for the woman in her conception and generation is but the imitation of the earth, and not the earth of the woman. (*Menexenus* 237d–238e)[5]

Derrida emphasizes Socrates's last remark that generation must be thought on the basis of the birth of the citizens of Athens from *khōra*, rather than from their biological mother, and thus in terms of autochthony. "A very important remark," he observes (1985–1986, "CHORA (suite)," 12). And he adds: "The proper of giving birth [*enfantement*] goes back to the earth, the earth gives the first example, the originary example of the originary [*originarité*], of the originary giving birth. Woman or the human mother imitated the earth, they learned from the earth what must be given to sons" (13). At this point, he recalls the analysis of autochthony developed by Loraux in *The Children of Athena*. This reference ends by evoking a remarkable passage from chapter 3, in which Loraux observes that, despite Socrates's emphasis on the generalized motherhood of *khōra*, the official representation of autochthony is not only patrilineal (*khōra* is *patris* too, in funeral orations [*epitaphioi logoi*]) but also aims to remove the biological process of sexual reproduction and generation. Before examining Loraux's text, I note that, throughout his work on *khōra*, Derrida highlights the inability of philosophy to speak about the mother that is beyond the father-son couple. This inability, I suggest, has to do with the fact that Derrida understands philosophy (namely, Platonism) as the system of autochthony, of the originary and nonmetaphorical relationship of the citizens of Athens with their fatherland.

In the pages that follow, I analyze the official representation of autochthony and *khōra* as it is discussed by Loraux in *The Children of Athena*. This allows us to shed light on Derrida's interpretation of Timaeus's recourse to *khōra* as a daring movement beyond the philosophical system of the logos-*zōon* and autochthony. In the introduction to her book, Loraux explains that she aims to develop "an Athenian topology" and thus to read "the myth [of autochthony (*auto-khthōn*, born from the soil of the fatherland)] *in* the city" by giving the "in" (1981, 8) a spatial value and thus linking the different versions of the myth to the civic spaces in which they take place. For instance, she recalls that, according to the *epitaphioi logoi*, all the Athenian citizens are granted an autochthonous origin from the soil of their fatherland, whereas, according to the attic ceramic, only Erichthonius, the founder of Athens, was born on the Acropolis from the earth fertilized by the sperm of Hephaestus.[6] From the outset, Loraux uncovers an interpretation of autochthony that, on my reading, Derrida had autonomously developed through his engagement with Plato's text, and that evidently captures his interest when he comes back to Plato in the eighties. Loraux supposes that autochthony, as it is elaborated in the *epitaphioi logoi*, allows the Athenians to suspend the biological process of generation and establish a nonmetaphorical relationship with their fatherland as with their origin, a relationship on the basis of which generation must be thought. Furthermore, she traces this concept of autochthony back to the aforementioned discourse of Socrates in Plato's *Menexenus*—in which women generate according to the example of the earth and not the opposite—by suggesting that we read it within the tradition of the *epitaphioi logoi*.

> First of all, how would Athenian men have responded to the question of their birth: were they born from the earth or from women? Will we have to admit that these citizens would have claimed that they actually sprung from earth? Such is indeed the response of the funeral oration, which, in its repeated references to the collective autochthony of Athenian citizens, dispossesses the women of Athens of their reproductive function. . . . For those who are still trying to determine a place for the maternity of woman in this whole affair, the *Menexenos* provides an easy solution—a bit too easy, in fact. "The woman" if we can believe the *epitaphios* given by Socrates, "imitates the earth." Generations of readers in search of a mother country called "Mother Earth" have believed that this affirmation could be simply lifted from its context, but in fact must be read in the framework of the Platonic *epitaphios*, where the signifier destroys the signified [translation modified]. . . . With this acknowledgement of Platonic irony, let us return to the conventional formulation of the *topos* in which women have no place whatsoever, while the earth, conspicuously called the *patris*, has as much to do with fathers as with the mother. (Loraux 1981, 9–10)

Loraux refers to the *Menexenus* once more in chapter 3, in the passage recalled by Derrida in the aforementioned 1985–1986 text. Here she argues that Plato's reference to the civic soil as a mother is quite isolated within the context of the orthodox representations of autochthony, according to which the civic soil has always already been reappropriated through the reference to the father made explicit in the recourse to the word *patris* (namely, fatherland). "The Attic earth is never only a 'mother' in the funeral orations," she observes, "but, through a return in full force of the paternal signifier, it is always both 'mother and father' (*mēter kai patris*: mother and earth of fathers)" (121).

In chapter 1, which Derrida invites his audience to reread carefully, Loraux advances the hypothesis of a topology of autochthony by which two places of the civic soil (namely, *khōra*), the Acropolis and the Ceramic, are identified as the sites of the two moments of the myth, the birth of Erichthonius and the inhumation of citizens, respectively. I have already quoted this text, according to which *khōra*, as mother and fatherland at once, is perfectly inscribed within the orthodox representation of autochthony. Autochthony allows the citizens of Athens to think the history of the city as the self-reproduction of the origin—that is, of the *aiōn*, which Loraux describes as "the vital principle always regenerated in the perpetual renewal of origins" (41). She explains:

> There is a short step between the immemorial and the timeless, a step soon taken by official eloquence, which suspends ordinary time in order to celebrate the *aiōn* of Athens. Here, generation after generation, (but is it even a question of successive generations?), the autochthonous citizens rejoin their common origin in the living present. Yet the immemorial is also the source of all nobility and thus "communal birth" (*koinē genesis*) becomes a synonym for "high birth" (*eugeneia*) in the funeral oration. As the central *topos* in the funeral oration, autochthony demands the general economy of the *logos*, which clarifies certain features of the discourse for us: its refusal of all genealogical considerations [this should make us think of the moment of Plato's *Sophist* in which the Stranger demarcates philosophy from genealogical discourses], for example, and, more importantly, the strange process by which the *polis* and the democracy escape all the laws of temporality. Even the notion of a history of the city's formation is foreign to the *epitaphioi*: as its own generating principle, Athens is instantly formed as Athens, and it is instantly political as well as civilized. Democracy is grafted onto autochthony. Who would recognize the history of Athens in all of this? And who would recognize the myth of Erichthonius?[7] (50)

Loraux further develops this reading of the Athenian concept of autochthony in her later work entitled *Born from the Earth. Myth and Politics in Athens* (1996), which I do not discuss here. To conclude this reading of Loraux, I remark, once more, that the system of autochthony described in *The Children of Athena* and, in particular, the representation of *khōra* as the civic soil of the fatherland, seem to merge with the zoological and political system that Derrida identifies as Platonism in

"Plato's Pharmacy," a system that I have described in my reading of this early text through the concept of the *logos spermatikos*.

THE BOLDNESS OF TIMAEUS

It is time to come back to Derrida's close reading of the *Timaeus* and to analyze why he interprets Plato's recourse to *khōra* as a breach opened up within Platonism and autochthony, and thus as the inauguration of another sequence of thought on genesis and history. I start this analysis by taking into account a series of texts from the 1970–1971 lecture course on *Conditions de l'inscription du texte de philosophie politique— l'exemple du matérialisme*.

As Derrida points out in a later note in *Khōra*, the course links "the reflections on the *Timaeus*" with the analysis of other texts, in particular Marx and Hegel, discussed "for the question of their relation with the politics of Plato in general, or the division of labor, or of myth, or of rhetoric, or of matter, etc." (1995, 149). I propose focusing on sessions 3, 4 and 6, in which we find a set of issues that Derrida further develops in later approaches to *khōra*: the nonphilosophical necessity of the recourse to *khōra*, *khōra* as a mother, the difference between the recourse to the third term in the *Timaeus* and the one undertaken by Plato in the *Sophist*, the double exteriority of *khōra*, the *mise en abyme* of the discourse on *khōra*.

In session 3, Derrida advances the hypothesis that Plato refers to *khōra* as to a pre-originary third term that remains external to the philosophical couple of the father and the son, identified as the intelligible paradigm and the sensible image.[8] The fact that Plato moves from the designation of *khōra* as a nurse (*pasēs einai geneseōs upodokhēn autēn oion tithēnēn*, *Timaeus* 49a) to that of *khōra* as a mother (*mēter*, 50d) does not affect *khōra*'s absolute externality to the field of philosophy. Derrida explains:

> However, it is evident that the mother, although she is no longer a nurse here, does not couple with the father, namely, with the model, the paradigm, as it was said earlier on, with insistence, that *khōra* is outside the couple of positions that the intelligible paradigm forms with sensible

> becoming, the father/son couple, so to speak. The mother is apart, she, *khōra,* is a place apart, spacing [*espacement*], as we will see, which entertains an always dissymmetric relation to that which, in or next to her, seems to couple with her. (1970–1971, 3.8–9)

Therefore, *khōra* (as a mother) is "pre-originary," "before and outside all generations" (3.9). It is for this reason that the discourse of Timaeus consists of backward movements that make him recommence each time. What is at stake, as Derrida puts it, is "going further, returning to what until now has served us as a beginning, going back on this side of principles, of the opposition between the paradigm and the copy" (3.9). Hence, the difficulties of Timaeus's discourse (bastard reasonings, metaphorical predicates, backward steps, etc.) mark the singularity of the research undertaken, "the boldness consisting in going back toward . . . what precedes and conditions the oppositions . . . that open up the assured discourse of philosophy" (3.10).[9] This pre-origin thus comes before the father-son couple as well as the philosophical discourse in general, which can deal only with this couple. I suggest that here Derrida understands Platonism as the system of the logos-zōon, whereas *khōra* (*qua* mother and pre-origin) accounts for the dissemination of the trace-seed. Therefore, he concludes his remarks as follows: "Philosophy as such can only speak about the father and the son as if the former engendered the latter by itself" (3.10).[10]

In session 4, Derrida emphasizes the originality of the step taken by Timaeus toward a pre-originary third term, by measuring this step against the Stranger's recourse to the third term (namely, Being) in the *Sophist*. Timaeus affirms the necessity of starting again the discourse about the origin of the universe by taking account of a third *genos* that is difficult and obscure (*Timaeus* 48e–49a).[11] The necessity of this leap beyond the philosophical couple is not justified, as Derrida suggests, because it is no longer (intra-)philosophical. "The necessity that commands and motivates this transition is not explained," he notes, "it does not give place to any philosophical justification, it is a leap beyond the philosophical opposition that raises the necessity of a third" (1970–1971, 4.7). Responding to the necessity of describing this third *genos* constitutes the engaging singularity of the text. Derrida highlights the "boldness" of this "transition [*passage*] to the third term" by recalling the analogous operation described by Plato in the *Sophist*, which he proposes understanding in the wake of Heidegger's interpretation. "In the *Sophist*," he explains, "the necessary recourse to the *triton ti* in 243d and 250b is, as you know, the very opening of the ontological question as such, of the question of Being"(4.7).[12] In this comparative reading, which is further elaborated in later writings, Derrida identifies the singularity of the *Timaeus* with the following two features: (a) the third of the *Sophist*, Being, has already been presupposed and employed by philosophical discourse, whereas the third of the *Timaeus* has not been anticipated and is irreducible to philosophical discourse (4.8);[13] (b) Being is still a *genos* or an idea, "it enters the community and the *symplokē* of ideas and genres" (4.8), it is grasped by the dialectical man, as Derrida notes, whereas *khōra* remains behind dialectics and breaks with it (4.9).[14]

The text that follows the step toward this pre-origin, according to Derrida, constitutes a "tropology" (*tropique*, 11), since *khōra* can be told only through inadequate analogies (such as those of the elements as well as of gold) and, ultimately, through a discourse that reproduces and mirrors itself in the very structure of *khōra* as receptacle, nurse, mother, etc. In order to respond to the necessity of unveiling the third term, Timaeus resorts first to the description of the nature of the elements and, then, to the analogy between *khōra* and gold.[15] Derrida observes that "the discursive qualifications of *khōra* are necessarily analogical, metaphoric, and inadequate. We are obliged to designate her by or according to the names of what she receives, bears, nourishes, etc." (4.10). Therefore, Timaeus proposes reconsidering the receptacle as a mother that, however, remains detached from the father as well as from the son:

> And it would not be out of place to compare the receptacle to a mother, the source to a father, and what they create between them to a child. We should also bear in mind that in order for there to exist, as a product of the moulding stuff, something that bears the whole multifarious range of visible qualities, the moulding stuff itself, in which the product is formed and originates, absolutely must lack all those characteristics which it is to receive from elsewhere, otherwise it could not perform its function. After all, if it were similar to any of the things that enter it, it would be no good at receiving and copying contrary or utterly different qualities when they enter it, because it would leave traces of its own appearance as well. (*Timaeus* 50d–e)

Khōra is a mother only to the extent that it allows for generation by remaining virgin. Derrida remarks on this passage as follows: "The receptacle remains always outside the *eidos* as well as outside any possible copulation with what represents the figure of the father . . . despite and beyond all impressions that the father will mark in her, inalterably virgin, inaccessible, intangible, and out of reach" (1970–1971, 4.14). Furthermore, this externality to copulation, this virginity, is what makes the variety of generations possible. Again, as Derrida puts it, "her perfect virginal indifference is the condition for [the] typomorphic richness" (4.14) to which she gives place. From this, Timaeus draws the implication that the receptacle must be external to all forms. He declares this externality twice, as Derrida remarks, by taking recourse to another analogy. "That is why," Timaeus argues by affirming the first externality, "if it is to be the receptacle of all kinds, it must be altogether characterless [*ektos eidōn*]" (*Timaeus* 50e). At this point, he evokes the example of the odorless liquids that receive the scents of perfumes and that of the uniform base stuff that welcomes the impressions of shapes.[16] Both are considered analogous to the receptacle of the copies of the intelligible paradigms. Here Timaeus affirms the second externality of *khōra* with respect to the *eidoi*: "The same goes, then, for that which repeatedly has to accept, over its whole extent, all the copies of all intelligible and eternally existing things: if it is to do this well, it should in itself be characterless [*ektos pantōn tōn eidōn*]" (51a). This double externality does not merely designate *khōra* but also prescribes the tropology that Timaeus's discourse on *khōra* seems to set in motion. I suggest that we read Derrida's remarks on this *ektōs tōn eidōn* from this twofold perspective:

> Therefore, polymorphy and proteomorphism suppose an intrinsic amorphism [*amorphie*], an aneidetic nature, and thus an irreducible non-figurability. Being absolutely figurable, the receptacle escapes all figures, it does not let itself be captured by any figure and necessarily exceeds the trope or the representation that are intended for it [*qu'on lui destine*]. (4.14)

Derrida further elaborates upon this seminal insight in his later, homonymous essay on *Khōra* where he suggests that not only Timaeus's discourse on *khōra* (and thus the tropology that makes up Plato's text) but the whole history of the interpretations of *khōra* is reflected and prescribed in *khōra* itself as the formless receptacle of the copies of intelligible forms. Therefore, *khōra* accounts for the very structure of this historical totality, for what allows us to conceive of the history of its interpretation as such. I refer to the last section of this chapter for an extended discussion of this hypothesis.

THE *DYNAMIS* OF *KHŌRA*

I observed that, in the 1985–1986 texts, Derrida brings his reading of the *Timaeus* into conversation with Loraux's reading of *khōra* as the civic soil of Athens. In this section, I draw attention to other relevant features that weld the text dated 22 January 1986 to the thinking of *khōra* that I have traced throughout these pages.

The text begins by affirming the absolute heterogeneity between the *khōra* of the *Timaeus* and that of the *Menexenus*—that is, between two understandings of genesis, the general structure of genesis and fatherland, dissemination and autochthony (including dialectical security), the trace-seed and the logos-*zōon*.[17] A few pages later, once again, Derrida brings to the fore the singularity of *Timaeus*'s *khōra* through the reference to the *Sophist*. As we know, he had already engaged in this comparative reading in the 1970–1971 seminar, in the wake of his analysis of the *Sophist* developed in the Heidegger course. The 1986 text focuses on the following three indexes: (a) *khōra* is "an eidos that does not resemble any other eidos" ("22 janvier," 4); (b) the discourse on *khōra* is "presupposed by any cosmogonic or anthropogenic story [*récit*] but is not an element of cosmogony or anthropogony," it is rather "the site where cosmogony and anthropogony take place" (4); (c) the "necessity [of the recourse to *khōra*] remains external to philosophy . . . or, at least, to what is designated as philosophy by Plato, namely, dialectics" (4). Furthermore, summarizing these indexes, the text points out that *khōra* is not inscribed within the *symplokē*, like every *genos*, and thus requires a leap beyond dialectics, which makes the security of dialectics and, more generally, of philosophy tremble. The following passage highlights the divergence between the transition to the third affirmed in the *Timaeus* and the discovery of the ontological question in the *Sophist*. The implications that follow from this divergence are further developed in *Khōra*, where Derrida rethinks history in light of *khōra* and, more precisely, he rethinks the latter as the very concept of history. The passage reads:

> In the *Timaeus*, the transition to the third does not take place within the community of an interlacing [*la communauté d'entrelacement*, like in the *Sophist*]. It is not a genre that circulates among other genres, like Being, a fifth or third genre among others. Here *Khōra*, and the insistence on this point is firm, is stranger to the couple of opposite terms and does not give place to any dialectics, thus to any philosophy. . . . Whereas Being—in the *Sophist*—inaugurated the tradition of the ontological logos [the logos-*zōon*?], the discontinuous transition to *khōra* seems to suspend the order and security of dialectical knowledge, which secures itself in the presence of the meaning [*sens*] of Being in language. Being is reassuring for language and the meaning of Being is what finally gives meaning to language, what is presupposed by language. Therefore, the thinking of the meaning of Being is what is indispensable for language to take place . . . [with *khōra*] we do not make explicit what we pre-comprehend in language. (1985–1986, "22 janvier," 6)

The text goes on with another reading of *Timaeus* 49b–50c. A series of original remarks on the analogy between gold and *khōra* follows. First, the text recalls that this analogy entails an absolute discontinuity so long as gold itself, being in turn the copy of an intelligible form, has already been impressed on *khōra*. Ultimately, this analogy uncovers the boldness of understanding *khōra* as what allows us to conceive of its inscriptions as a totality, and thus as the structural law of this totality. "The 'like' [*comme*] illustrates well the analogy," we read, "but it illustrates also the absolute shift, the rupture of homogeneity because gold itself is in *khōra*." The text continues: "Gold is a figure of *khōra* as she can receive it," and thus *khōra* is "non-figurable" (10).[18] *Khōra* is not the receptacle of its *own* determinations—as the mother would be, properly—and it could not be otherwise, I suggest, as it provides the minimal condition for generation in general and thus for the totality of generations. The text describes the "paradoxical logic" that undergirds this understanding of *khōra*:

> *Khōra* is neither the support nor the receptacle of the qualities that would belong to her on her own [*en propre*], she does neither support nor receive by herself [*en propre*]. Therefore, she does neither support nor receive anything.

She is not what she is, what she supports, the forms she takes on; she does not let herself be affected by any determination precisely because she must be able to receive all determinations. (11)

This logic necessarily affects the task of unveiling (*emphanisis*), whose nonphilosophical (or more-than-philosophical) necessity is conjured up in *Timaeus* 49a.[19] We refer to *khōra* only by means of the names that it *can* receive (see the example of gold), names that figure *khōra* by covering and dissimulating it.[20] This is what is called "the insurmountable paradox of the *emphanisis* of *khōra*" ("22 janvier," 11), the fact that every interpretation of *khōra* is a tropology and—to anticipate Derrida's later determination—an anachronism. Commenting on *Timaeus* 50d, a few pages later, the text suggests that the *dynamis* of *khōra*—this is the word used by Timaeus when he wonders about the natural property of the third term (49a)—consists in a sort of principle of indetermination that is an unconditional opening to the to-come. *Khōra* must be absolutely undetermined in order to be absolutely receptive. As we will see, *khōra*'s *dynamis* accounts for the generality of generation and the concept of history. Therefore, we must acknowledge that *khōra* does not keep any of the forms received, otherwise:

> It [the receptacle] would oppose a resistance where it must be absolutely receptive. Its virginal indifference allows it to give place—we should say, to lend place—to the greatest typomorphic diversity . . . in order to be polymorphic as well as proteiform, *khōra* must remain intrinsically amorphous, somehow aneidetic and non-figurable (42–43).[21]

These remarks are followed by the analyses of the double externality of *khōra* (*Timaeus* 50e–51a), of the ways of access to the thought of *khōra* (52b), and, finally, of Aristotle's interpretation of *khōra* as matter and of the repercussions that this interpretation produces on the history of the interpretations of *Timaeus*'s third term.

THE CONCEPT OF HISTORY

In this last section, I undertake an examination of Derrida's late essay *Khōra* according to the coordinates of reading that I have established in this chapter. In particular, I draw attention to those features that Derrida had discussed in the previous approaches to *khōra* and that he further elaborates in this text, the latest one explicitly dedicated to the topic.[22] *Khōra* was originally included in a collection of essays in homage to Jean-Pierre Vernant in 1987. This is testified by the exergue and the initial section, in which Derrida directly addresses the French scholar of Greek thought.[23] A more or less implicit reference to Vernant and his work on the relationship between myth and logos can already be found in the earlier text on *khōra* dated 8 January 1986, which identifies its task with the analysis of *khōra*'s irreducibility to the philosophical logos as well as to the mythological one.[24]

In *Khōra*—this is my hypothesis—Derrida pushes his reading of the *Timaeus* to its limits by unfolding its implications for our understanding of history. He shows that the thinking of genesis that underlies the movement toward *khōra* solicits the understanding of history presupposed by Platonism and highlights the structural law and concept of history—that is, the minimal condition that allows us to think of history as such.

Derrida's *Khōra* demarcates itself within the history of the interpretations of *khōra*. It represents the boldness of Derrida's reading of *khōra*, in the wake of and somehow further than the boldness of Timaeus's leap beyond Platonism. In *Khōra*, Derrida demonstrates that all the interpretations of *khōra* are interpretations precisely because of the very structure of *khōra*, and that their being as a historical totality constitutes an effect of this structure. First, he addresses the structural law by which "tropology and anachronism are unavoidable" (1995, 94) for every discourse on *khōra*. He points out that a discourse on *khōra* is bound to impress on *khōra* itself a determination that remains stranger to it:

> Whether they concern the word *khōra* itself ("place," "location," "region," "country"), or what the tradition calls the figures proposed by Timaeus ("mother," "nurse," "receptacle," "imprint-bearer"), the translations remain caught in the network of interpretation. They are led astray by retrospective projections, which can always be suspected of being anachronistic. (93)

For the first time in the history of the interpretations of *khōra*, Derrida's text explains that the irreducibility of anachronistic retrospections constitutes a structural effect of *khōra* itself—as the receptacle of impressions and determinations—and thus of the process of genesis that *khōra* initiates. "And all we would like to show," Derrida writes "is that it is structure which makes them [tropology and anachronism] inevitable . . . it is this structural law which seems to me never to have been approached as such by the whole history of the interpretations of the *Timaeus*" (94). The immediate consequence of this hypothesis is that we can think of the historical totality of the interpretations of *khōra* as such, as a totality—namely, a sharing of prescribed features—that is indefinitely open to interpretations to come. The process of genesis initiated by *khōra* is what makes possible to think this historical totality, the open totality of its interpretations. From this, a second hypothesis follows: "The presumption of such an order (grouping, unity, totality, organized by a *telos*) has an essential link with the structural anachronism of which we spoke a moment ago. It would be the inevitable effect produced by *something like khōra*" (94).

What I have presupposed and left implicit so far is the structure of *khōra*, whose effects are the interpretative status of all discourses on it and their interminable history. As Derrida points out, all discourses on *khōra* reproduced, reproduce, and will reproduce what Timaeus says about *khōra* and thus what is announced in the very structure of *khōra*: the fact that it is the non-figurable receptacle of all figurations. Therefore, the very discourse about a discourse on *khōra* always takes up the tropological system that has been used to determine *khōra* from the outset—that is, since the bold interpretation of *khōra* given by Timaeus in Plato's text. Derrida writes:

> Rich, numerous, inexhaustible, the interpretations come, in short, to give form to the meaning of *khōra*. They always consist in *giving form* to it by determining it, it which however can "offer itself" or promise itself only by removing itself from any determinations, from all the marks or impressions to which we say it is exposed. . . . But what we are putting forward here of the interpretations of the *khōra*—of Plato's text on the *khōra*—by speaking about a form given or received, about mark or impression, about knowledge as information, etc., all of that already draws on what the text itself says about *khōra*, draws on its conceptual and hermeneutic apparatus. (1995, 94–95)

Through this new interpretation of *khōra*, Derrida sheds light on a non-Platonic understanding of history. Timaeus's discourse on *khōra* programs and prescribes the shared features of all the interpretations of *khōra*, the fact that they are interpretations and thus tropologies and anachronisms. However, this program does not neutralize the totality of interpretations, as if it were a spectacle without duration before an incorruptible spectator. Here we touch upon the implications of the opening onto *khōra* for our understanding of history. The structure of *khōra* constitutes the minimal condition for history as a totality of interminable interpretations, a totality that has never been given in the present and remains open to the to-come. This structure accounts for what can be programmed and prescribed of history—that is, the very concept of history.

> *Everything happens as if* the yet to come history of the interpretations of *khōra* were written or even prescribed in advance, *in advance reproduced and reflected* in a few pages of the *Timaeus* "on the subject" of *khōra* "herself" ("itself"). . . . Is a prescribed history, programmed, reproductive, reflexive history still a history? Unless the concept of history bears within itself this teleological programming which annuls while constituting it. (99)[25]

Derrida provides us with another perspective from which we can look at what Timaeus does by taking his step back toward *khōra*. If *khōra* bears within itself the program and prescription of the history of its interpretations, by speaking about *khōra* Timaeus uncovers an abyss in the middle of the text, an abyss in which the very structure and fate of his discourse—as well as of the history that includes this discourse—are explained. The boldness of Timaeus may be reconsidered in relation to this abyss—namely, to the (pre-)origin of his discourse as well as of history. "Didn't it name a gaping opening, an abyss or a chasm?" Derrida wonders. And he goes on, "Isn't it starting out in this chasm, 'in' it, that the cleavage between the sensible and the ideal . . . can have place?" (103). Therefore, *mise en abyme* is the structure of *khōra*, which brings about the interminable series of its interpretations.[26] It is what can be programmed and prescribed of history, what makes us think of history as such, of the concept and historicity of history. As Derrida puts it, "A *mise en abyme* regulates a certain order of the composition of discourse" (104). This structure prescribes what happens to the inscriptions on *khōra* and, more generally, to the disseminated trace-seed that takes place at the origin of the world. "*Mise en abyme* of the discourse on *khōra*," Derrida writes, ". . . such would be, then, the structure of an overprinting without a base" (104).

At this point, it is worth interweaving this interpretation of the relationship between *khōra* and the totality of its interpretations and our reading of the breach opened up within Platonism by the movement toward *khōra*. The structural law of *khōra* and its implications for a certain understanding of history presuppose an understanding of genesis that dissociates the latter from Platonism, autochthony, Athens, etc. In the concluding pages of the text, Derrida comes back to Timaeus's determination of *khōra* as a mother (*Timaeus* 50d). He recalls that *khōra* is a mother "apart," or a "strange mother," to the extent that it accounts for another concept of genesis and engendering, a nonanthropomorphic one, which is designated as a giving place. Here, I argue, Derrida focuses on the shift between the kind of genesis that *khōra* makes possible and an anthropomorphic understanding of genesis according to which the engendering of the mother is derived from the father-son relationship and thus from the logos-*zōon* of Platonism. Derrida writes:

> The mother is "supposedly" apart. And since it is only a figure, a schema, therefore, one of these determinations which *khōra* receives, *khōra* is *not* more of a mother than a nurse, is no more than a woman. This *triton genos* is not a *genos*, first of all because it is a unique individual. She does not belong to the "race of women" (*genos gynaikōn*). *Khōra* marks a place apart, the spacing which keeps a dissymmetrical relation to all that which, "in herself," beside or in addition to herself, seems to make a couple with her. In the couple outside of the couple, this strange mother who gives place without engendering can no longer be considered as an origin. She/it eludes all anthropological schemes, all history, all revelation and all truth. Preoriginary, *before* and outside all generation, she no longer even has the meaning of a past, of a present that is past. Before signifies no temporal anteriority. The relation of independence, the nonrelation, looks more like the relation of the interval, or the spacing to what is lodged in it to be received in it. (1995, 124–25)

The understanding of genesis that grounds Platonism and is embodied by the autochthonous community of Athens is called philosophy. Its element is the logos-*zōon* (and thus the originary and nonmetaphorical relationship between the father and the son). *Khōra* withdraws from the field of philosophy, as a meta-philosophical necessity that remains unheard-of or is removed, that bears within itself another thinking of history, the only thinking of the concept and historicity of history. Philosophical language is not only unable to address the structure of *khōra* as such. Above all, it thinks of engendering in light of the relationship between the logos and its father-subject and thus through an anthropomorphic concept of engendering as the self-reproduction and incorruptibility of the same. In conclusion, Derrida accounts for the gap between *khōra* and philosophy by reproducing a formula that he had used in his previous texts on *khōra*: "Philosophy cannot speak *philosophically* [my emphasis] of that which looks like its 'mother,' its 'nurse,' its 'receptacle,' or its 'imprint-bearer.' As such, it speaks only of the father and the son as if the father engendered it all on his own" (126).[27]

3

HEGELIANISM I
Tropic Movements

> She puts into practice—and thus to the test—this other Hegelian motif: the very possibility, or chance, of a *speculative* language, or rather, of inserting the *speculative* in language when language measures itself to the very capacity of condensing two contradictory meanings in one single syntagm. Two antinomical meanings *at the same time* concentrated in one and the same verbal formation.... The Hegelian *Aufhebung* is not only an example of this possibility: it is its very concept.
>
> —Derrida, "A time for farewells: Heidegger (read by) Hegel (read by) Malabou," 11

Philosophical discourse, such as the logos-*zōon* and, more generally, the *logos spermatikos*, constitutes the element of the organization of ontological regions and discourses as well as of the metaphorical exchanges among them. As I have attempted to demonstrate in the two previous chapters dedicated to Derrida's reading of Plato's text, this is what the French philosopher interprets as the thesis of Platonism. He points out that Platonism wishes to destroy its textuality—namely, the anagrammatic structure of the trace-seed and the nonphilosophical necessity of *khōra*, in view of demarcating itself as philosophical discourse. In the years subsequent to "Plato's Pharmacy," Derrida engages in the project of uncovering the organization of regional discourses and thus the relationship between philosophy and the natural sciences (especially biology) within the Hegelian text, which he calls "Hegelianism," the Hegelian "system," or the "book of life" (as we see later). In this chapter and the next ones, I examine how Derrida carries out this project by drawing together unpublished as well as published and well-known texts. In particular, I focus on session 1 in the unedited lecture course on *Théorie du discours philosophique: La métaphore dans le texte philosophique* (*Theory of Philosophical Discourse: The Metaphor in the Philosophical Text*, 1969–1970), the preface to *Dissemination* (1972, "Outwork, Prefacing"), the texts collected in *Margins of Philosophy* (1972, especially "White Mythology: Metaphor in the Text of Philosophy," already included in the aforementioned seminar and originally published in *Poétique* in 1971) and *Glas* (1974, whose left-hand column corresponds to the revised text of the unedited course on *La Famille de Hegel* [*Hegel's Family*] taught by Derrida in 1971–1972).

THE PHILOSOPHICAL INTROJECTION OF ORDINARY LANGUAGE

The opening session of the aforementioned 1969–1970 lecture course begins with a programmatic note that, in an explicit fashion, ties the analyses that follow to the earlier reading of Platonism. "The implicit project" of the seminar, Derrida points out, "is calling into question what constitutes somehow the essence and telos of philosophy, that is, holding [tenir] the most general discourse, and thus the most independent one, in relation to which particular discourses (determinate domains) would be hierarchically ordered" (1.2). Derrida recalls the predicates that Aristotle had assigned to philosophical discourse understood as metaphysics (or *philosophia prōtē*)—namely, priority and universality. From within the Aristotelian framework, he raises, once more, the question about the hierarchy and order of metaphorical exchanges between regional discourses (especially, between philosophical and scientific discourse).[1] He remarks that "This double character of priority and universality, which will always belong to the representation philosophical discourse grants itself, entails that philosophical discourse, as the discourse of or about Being, can never fall under the jurisdiction of a particular science, of a discourse that refers to a determinate type of being" (1.3). As I have already pointed out, this is a key question for Derrida as it concerns the relationship between linguistics and zoology and thus the minimal structure of the logos as well as of the living.

The whole burden of the question rests on the philosophical concept of metaphor, which, by definition, weaves together two regional discourses within a system. In the passage from "Plato's Pharmacy" where he discusses the implications of Plato's paternal thesis, Derrida explains that, within the limits of Platonism, family is not a natural and biological concept but rather an effect of the originary and nonmetaphorical relationship between the logos-*zōon* and its father, which thus constitutes the element of metaphorical exchanges. Therefore, it is not by chance that, in the 1969–1970 lecture course, he takes up the philosophical discourse as the most domestic discourse, that is, the most familiar one, family itself, and thus as the *logos spermatikos*.[2]

The question already posed in "Plato's Pharmacy"—what is the father?—comes back here. Is the father a figure of the natural and biological genitor (such as the genitors of the Hegelian philosophy of nature) or, rather, the origin and power of the logos? How does philosophical discourse explain the introjection, exploitation and, finally, appropriation of biological and, more generally, scientific images?

> At the very moment when philosophical discourse believes it is the most autonomous discourse, when its autoregulation appears as the most assured and domestic one, precisely then, it must take into account [*compter avec*] heterogeneous elements, of different origins and kinds, different for structure and functioning: this is what will interest us above all. It is the system of this heterogeneity, of these models of importation, or if you prefer of these importations of models, that we must try to master [*dominer*] and start formalizing. Therefore, I say by way of anticipation and quite directly that the concept of metaphor and metaphorization can designate at once, in an equivocal fashion, one of these models of importation, a form among others of these borrowings or of this trafficking [*de cette contrebande*], and the general form of the introjection of an allogenic element into philosophical discourse. Metaphorization can be understood, strictly speaking, as philosophy's borrowing of a particular figure from ordinary language or, broadly speaking, as the general form

of displacement by which philosophy exploits a discourse that is stranger to it, by borrowing or assimilating, in one way or another, lexicon, syntax, content, result, etc. (1.5)

It is worth remarking that here Derrida has recourse to the biological image of autoregulation in order to account for the relationship between philosophical discourse and external and allogenic factors. I advance the hypothesis that this programmatic passage aims to call into question not only the primacy of philosophical discourse above the biological one but also the thesis that, by consequence, biological autoregulation and, more generally, the structure of the living, should be thought on the basis of philosophical discourse (namely, father-son relationship, family, the logos-*zōon*, the *logos spermatikos*, etc.). The interrogation that I examine in the following sections argues for a non-philosophical understanding of biological autoregulation and of the living in general, an understanding that, since my reading of "Plato's Pharmacy," I have described as the dissemination of the trace-seed.[3]

THE LIFE OF THE CONCEPT

In the subsequent pages of the course, Derrida takes into account the case of the Hegelian analyses of the relationship between philosophy and the sciences. He determines this relationship as one of "fascinated resistance" (1.5), thus suggesting that philosophy cannot help resorting to what it wishes to keep outside itself and, therefore, it makes itself afraid.[4] To illustrate how Hegel explains this relationship, Derrida comments on his own translation of the following passage from *Science of Logic* (1812–1831) section I.1.2.2:

> Essentially, however, the *perversity* [my emphasis] of enlisting mathematical categories for injecting some determination into the method and the content of philosophical science shows in the fact that, inasmuch as mathematical formulas signify thoughts and conceptual distinctions, this meaning must rather first be indicated, determined and justified in philosophy. In its concrete sciences, philosophy must take its logical element from logic, not from mathematics; it can only be an expedient of philosophical incapacity to resort

for the logical element of philosophy to the shapes which it assumes in other sciences, many of which are only adumbrations of this element and others even perversions of it. Besides, the mere application of such borrowed formulas is an external operation; the application itself must be preceded by the awareness of both their value and their meaning, and only the consideration of thought, not any authority drawn from mathematics, yields this awareness. Logic itself is this awareness regarding such formulas. It strips them of their particular form, rendering it superfluous and useless; it rectifies them and alone procures for them their justification, their sense and value. (Hegel 2010, 181)[5]

A parenthetic remark follows the translation of Hegel's text, in which Derrida draws attention to what in the quoted edition of the *Science of Logic* is notably translated by "perversity." He observes that the original German *als etwas verkehrtes* literally means "like something inverted, overturned, taken in the wrong way [*comme quelque chose d'inverti, de renversé, pris à contre-sens*]." "Wanting to legislate in philosophy on the basis of a determinate science," he continues, "is properly a non-sense [*contre-sens*], it goes against sense or in the wrong sense [*en sens inverse*], it overturns the order of sense, and so on" (Derrida 1969–1970, 1.6). Therefore, in the wake of the already discussed reading of the logos-*zōon* in "Plato's Pharmacy," the bearing of this passage hinges on the order, the orientation, the arrow, the way—namely, the *sense*, of the relationship between philosophy and the natural sciences (and thus of the metaphorical exchanges between them). Derrida seems to suggest, once again, that regional discourses are grounded on the philosophical presupposition and thus, as he concludes a few paragraphs below, "the sense [*sens*] of sciences, this is philosophy [*c'est la philosophie*]" (1.7). According to Derrida, this sense, which organizes the relations among regional discourses and secures the primacy and universality of philosophy, consists in a concept of life—the life of the concept or, to play with Plato's logos-*zōon*, concept-life—on the basis of which regional discourses, such as mathematics or biology, must be understood. This concept is inscribed within the philosophical tradition of the *logos spermatikos*, which, according to Derrida, understands life as the generation of consciousness and thus as the nonmetaphorical and originary relation between the father and the logos-*zōon*, as the removal of the mother, self-reproduction, etc. Derrida formalizes the element of philosophical discourse as follows:

> However, it is worth remarking that the critique of formalism and mathematicism refers, as to its ultimate foundation, to a concrete life, to being as present life, the life of the concept in Hegel and that of the transcendental consciousness in Husserl. In both cases, we cannot dissociate from this life, from this liveliness [*vivance*] (*Lebendigkeit*), the concept of consciousness. As you know, it is in the name of the life of the concept that Hegel criticizes the empty formalism and the cadaveric rigidity of understanding [*entendement*] and mathematical schemes as well as the abstract and empty transcendentalism of Kant. (1.7)

The passage explains why, as anticipated, calling the Aristotelian primacy and presupposition of philosophical discourse into question is for Derrida intimately linked with the liberation of a non-philosophical discourse about life (namely, a certain bio-logy), such as the thinking of dissemination.

The most basic feature of the organization of regional discourses concerns the relation between philosophy and natural language. For Derrida, Hegel conceives of natural language by holding on to the Aristotelian presupposition that he has discussed so far. For instance, philosophical discourse can draw resources from the German language so long as it finds in the latter, more precisely, in some German words, the expression of philosophical concepts and thus speculative spirit. "In the eyes of Hegel, there are languages that are *naturally speculative* [*naturellement spéculatives,* my emphasis]," Derrida explains, "naturally more or less speculative, more or less appropriate to name [*nommer*] speculative, that is, dialectical concepts, which encompass within themselves contradiction and the unity of contradiction, the identity of identity and non-identity" (1.9). This passage refers implicitly to the second Preface of the *Science of Logic*, which was added to a later edition (1831). This preface takes up the observations that Hegel had already developed in another text from the same work, the remark on the use of the term *Aufhebung* that had been included in *Science of Logic* section I.1.1.1 since its first edition (1812–1813). As we see below, Derrida had already commented on this text in "Violence and

Metaphysics" (originally published in *Revue de métaphysique et morale*, in 1964). Furthermore, Jean-Luc Nancy refers to it as to the point of departure of the tour de force mobilized in *The Speculative Remark: One of Hegel's Bon Mots* (1973), a tour de force that is evidently inspired by Derrida's reading of the same text. Derrida translates the following passage from Hegel's preface:

> It is to the advantage of a language when it possesses a wealth of logical expressions, that is, distinctive expressions specifically set aside for thought determinations. Many of the prepositions and articles already pertain to relations based on thought (in this the Chinese language has apparently not advanced that far culturally, or at least not far enough), but such particles play a totally subordinate role, only slightly more independent than that of prefixes and suffixes, inflections, and the like. Much more important is that in a language the categories should be expressed as substantives and verbs, and thus be stamped into objective form. In this respect, the German language has many advantages over other modern languages, for many of its words also have the further peculiarity of carrying, not just different meanings, but opposite ones, and in this one cannot fail to recognize the language's speculative spirit. It can delight thought to come across such words, and to discover in naive form, already in the lexicon as one word of opposite meanings, that union of opposites which is the result of speculation but to the understanding is nonsensical. (Hegel 2010, 12)

As Derrida points out, the fact that some German words draw together two opposed meanings (*entgegensetzte*) in their graphematic unity is "absurd as contradictory [*contradictoire*] (*widersinnig*)" for formal understanding. At the same time, it is "the result of speculation, of what Hegel calls speculative, dialectical thought, which overcomes formalism as well as empiricism" (1969–1970, 1.10).[6] Hegel goes on by explaining that, therefore, "philosophy stands in no need of special terminology . . . where everything depends on meaning the most" (Hegel 2010, 12).[7] He points to "the advance of culture and of the sciences in particular," which "fosters the rise of thought-relations that are also more advanced" (13), as the principle of the systematic organization of languages and discourses.[8] It is on account of this principle that we can follow the traces of speculative spirit also in scientific categories, such as that of polarity and of the kind of difference it describes. Hegel continues:

> In Physics, for instance, where the predominant category previously was that of force, it is the category of polarity that now plays the most significant role—a category which, incidentally, is randomly being imposed all too often on everything, even on light. It defines a difference in which the different terms are inseparably bound together, and it is indeed of infinite importance that an advance has thereby been made beyond the abstractive form of identity, by which a determinateness such as for example that of force acquires independent status, and the determining form of difference, the difference that at the same time remains an inseparable. (13)

This passage displays in an explicit fashion the key issue for the interrogation of Hegelianism that Derrida takes up in his unedited course: the fact that the same term can be used in two different discourses and that philosophy constitutes the sense of the exchange between these discourses (namely, a certain concept of life). Conjuring up the example of the physical category of polarity, Derrida concludes that "the same concept or, this is the problem, the same word is invested by a certain value, activity, or operativity, in two fields at once" (Derrida 1969–1970, 1.12). A question follows from this: "What is the status of this *homology* [my emphasis]?" (1.12), a homology—namely, an intersection of discourses—that, as we see below and in the next chapter, is also designated by Derrida as equivocity, double mark, and remark.[9]

THE HEGELIAN TREATMENT OF EQUIVOCITY

I pause with the examination of the lecture course for a moment and take a few steps backwards in order to trace Derrida's earlier analyses of the philosophical explanation of homology. I observed that Derrida had drawn attention to the Hegelian note on the speculative spirit of the German language by analyzing the remark in *Science of Logic* section I.1.1.1. I referred to part 1 of "Violence and Metaphysics," where Derrida focuses on Emmanuel Levinas's thought of absolute exteriority as a nonspatial exteriority. Levinas seems to make himself afraid by obliterating the spatial inscription (namely, "syntax") that allows him to conceive of exteriority.[10] Holding on to Leibniz's thesis of the unity between civil and philosophical language,[11] Derrida argues that we cannot neutralize the exchange and equivocity between the two languages but we should reckon with them, precisely as Hegel does with the German language.

> No philosophical language will ever be able to reduce the naturality of a spatial praxis in language; and one would have to meditate the unity of Leibniz's distinction between "civil language" and "scholarly" or philosophical language. And here one would have to meditate even more patiently the irreducible complicity, despite all of the philosopher's rhetorical efforts, between everyday language and philosophical language; or, better, the complicity between certain historical languages and philosophical language. A certain ineradicable naturality, a certain original naïveté of philosophical language could be verified for each speculative concept (except, of course, for the nonconcepts which are the name of *God* and the verb *to be*). Philosophical language belongs to a system of language(s). Thereby, its nonspeculative ancestry always brings a certain equivocality into speculation. Since this equivocality is original and irreducible, perhaps philosophy must adopt it, think it and be thought in it, must accommodate duplicity and difference within speculation, within the very purity of philosophical meaning. No one, it seems to us, has attempted this more profoundly than Hegel. (Derrida 1978, 141–42)

What does "adopting" this equivocity of language mean? What does Hegel exactly do, according to Derrida? In the subsequent passage, Derrida seems to refer to Hegel's case as exemplary of a certain engagement with the irreducible equivocity of philosophical language. However, he does not distance himself from the Hegelian treatment of equivocity, which, as it appears evident later, is based on the philosophical presupposition by which speculative concepts are the sense of natural language. "Without naïvely using the category of chance, of happy predestination or of the chance encounter," Derrida limits himself to observing, "one would have to do for each concept what Hegel does for the German notion of *Aufhebung*, whose equivocality and presence in the German language he calls *delightful*" (142). In the later text entitled "White Mythology," Derrida comes back to the Hegelian solution of the equivocity between natural and philosophical language by referring to the analogous case of the German word *Sinn* (namely, sensible/intelligible sense). On this occasion, he marks the distance between the delight of speculative thought and the interrogation of the anagrammatic, tropic and syntactical structure of language (as well as of the living).

Indeed, Derrida had engaged with this question of equivocity even earlier: since his work on Husserl, in the pages of the *Introduction* to the *Origin of Geometry* where he discusses the origin and use of language in transcendental phenomenology (section 5). In a footnote commenting on *Ideas I* section 59, in which Husserl opts for the irreducibility of logical axioms (such as the principle of noncontradiction) in pure consciousness, Derrida writes:

> But he says nothing about the language of this ultimate science of pure consciousness, about the language which at least seems to suppose the sphere of formal logic that we just excluded. For Husserl, the univocity of expression and certain precautions taken *within* and *with the help* of language itself (distinctions, quotation marks, neologisms, revaluation and reactivation of old words, and so on) will always be sufficient guarantees of rigor and non-worldliness. That is why, despite the remarkable analyses which are devoted to it, despite the constant interest it bears (from the *Logical Investigations* to the *Origin*), the specific problem of language—its origin and its usage in a transcendental phenomenology—has always been excluded or deferred. (Derrida 1989, 68)

Therefore, Derrida finds in language, understood as what precedes the distinction between the natural and the transcendental, the inescapable problem of transcendental phenomenology. "To the very extent that language is not 'natural,'" he argues, "it [language] paradoxically offers the most dangerous resistance to the phenomenological reduction, and transcendental discourse will remain the most irreducible difficulty" (62–63). He recalls that Eugen Fink had drawn attention to this problem in the seminal paper presented at the Husserl conference in Royaumont in 1957 and entitled "*Les concepts opératoires dans la phénoménologie de Husserl*" ("Operative Concepts in Husserl's Phenomenology"). In this paper, Derrida remarks, Fink attributes "a certain *equivocation* [*une certaine équivoque*, my emphasis] in the usage of operative concepts (that of 'constitution,' for example) . . . to the fact that 'Husserl does not pose the problem of a transcendental language'" (69). Furthermore, Derrida refers to the suggestive case of Suzanne Bachelard who, with regard to the expression "intentional life," "evokes the danger of 'a surreptitious return to psychologism' for 'language does not know the phenomenological reduction and so holds us in the natural attitude'" (69). What is at stake here is an earliest reference to the idea of life that, according to session 1 in the 1969–1970 lecture course, inscribes Husserl's work within a certain tradition of philosophy as the sense of sciences and thus as the *logos spermatikos*.[12] In his *Introduction*, Derrida focuses on the equivocity of language and pushes it to its limits by pointing to a minimal structure (or an element) that would be diffracted into the transcendental signification and the natural one, and thus would remain behind their difference. This structure is what he calls the anagrammatic, tropic or syntactical dimension of language, the textuality of a text. He writes:

> On the basis of the problems in the *Origin*, we can thus go on to ask ourselves, for example, what is the hidden sense, the nonthematic and dogmatically received sense of the word "history" or of the word "origin"—a sense which, as the common focus of these significations, permits us to distinguish between factual "history" and intentional "history," between "origin" in the ordinary sense and phenomenological "origin" and so on. What is the unitary ground [*fondement unitaire*] starting from which this diffraction of sense is permitted and intelligible? What is *history*, what is the *origin*, about which we can say that we must understand them sometimes in one sense, sometimes in another? (Derrida 1989, 69)

Finally, for the first time, Derrida recalls Althusser's recent translation of Feuerbach's essay *Towards a Critique of Hegel's Philosophy*. He finds in this essay the instance of a radical critique of the concept of origin that would refrain all readers from "the mythology of the absolute beginning" (69). I will further develop these remarks in the following pages through the analysis of "White Mythology," where Derrida explicitly demarcates himself from the Hegelian and, more generally, philosophical treatment of equivocity, and engages in the search for something like the unitary ground that he had highlighted in the earlier text of the *Introduction*.

Returning to the 1969–1970 course, I take up once more the question of homology that Derrida had raised in relation to the enjoyment of thought before the apparition of speculative spirit in the German language. According to Derrida, philosophy has always already avoided this question, precisely as philosophy consists in the sense of sciences and this sense amounts to a certain concept of life (such as the logos-*zōon*, the life of the concept, and transcendental life). "Philosophy is the project of a general organization of discourses governed by the legislation of an ultimate instance [*par une législation de dernière instance*]," he remarks, "philosophy is constituted by this project in a substantial, essential fashion" (1969–1970, 1.14). In the lecture course, Derrida focuses on Hegel's take on the equivocity of *life* by sketching out a reading that he further elaborates in the later seminar on *La Famille de Hegel* and, subsequently, in *Glas*. For Derrida, this take is exemplary to the extent that it accounts for the relationship between philosophy and the sciences and thus between the life of the concept and biological and natural life. Referring to *Science of Logic* section II.3 (entitled "The Idea"), Derrida explains that the proper of the Hegelian concept is life, not natural but spiritual life, life as the natural life negated and conserved, in one word, sublated (*relevée*) through the process of *Aufhebung*, and thus as life that sublates itself. Therefore, according to Derrida's reading of Hegelianism, the equivocity of life and that of *Aufhebung*, which constitute the leading thread of the investigation conducted in *La Famille de Hegel* and in *Glas*, are irreducibly interwoven. The course reads:

> Absolute Logos, the concept in Hegelian terms, has this essential character, attribute or property, that it is living [*vivant*]. I refer you to the last section of the *Logic of the Idea*, among other texts, whose first chapter concerns Life (it says that "the idea is in the first place life") and whose

last chapter, *The Absolute Idea*, posits the dialectical identity of life and being: "the absolute idea is Being, imperishable Life (*unvergänglichen Leben*), truth that knows itself, it is entirely truth." This equivalence of idea, being, truth, and life, is certainly not immediate; and, somehow, the immediacy of the determination of the idea in life is further negated, but negated and conserved, sublated [*relevée*] according to the contradictory movement of *Aufhebung*. Here we have to do with this double rooting [*enracinement*] in a natural-speculative language and in metaphoricity, which appeals, on one hand, to this untranslatable concept of *Aufhebung* and, on the other, to the figure of the circle that we have entered today and we are not going to leave soon. Indeed, if the immediate determination of the absolute idea as life is negated, it is negated as natural life and sublated as spiritual life; it is with the equivocity [*l'équivoque*] of this value of life that we have to do. (1.13–14)[13]

As Derrida suggests here, Hegel takes on the equivocity of life as well as of *Aufhebung* by presupposing the speculative concepts of life and *Aufhebung*. The solution of this equivocity consists in the very sense and organization of the relations among regional discourses. Therefore, natural and biological life, the life of scientific discourse, is thought on the basis of spiritual life—that is, of the life of the concept (or the concept-life). Derrida explicitly evokes a sort of precomprehension:

> The problem that is once more displayed and exemplified here is the problem of the relationship between the history of philosophical discourse and that of scientific discourse, for example, biological discourse. In Hegel, this problem is somehow taken over [*relayé*] by the problem of the relations between logic and the philosophy of spirit, on the one hand, and the philosophy of nature, on the other hand. Out of which linguistic evidence, out of which ordinary pre-comprehension, do we speak about life in biology, in the philosophy of nature, and in the philosophy of spirit? All problems of articulation among empiricity, philosophy, and the sciences, go through [*passent par*] this question. (1.15)

This precomprehension—that is, the philosophical presupposition of the sense and organization of regional discourses, and thus the principle of concept-life—is found at work in the Hegelian concept of the philosophy of nature, which positions scientific discourse in relation to the philosophical one (namely, logic and the philosophy of spirit). Derrida understands the relationship between the philosophy of nature and the philosophy of spirit and the transition from one to the other, within the Hegelian system, as the key threshold of his interrogation of Hegelianism and of his thought of dissemination. The next chapter puts this hypothesis to the test by reinterpreting the Hegelian system according to the coordinates of reading established by Derrida in the 1969–1970 course—namely, as the book of life. The aforementioned remarks on the sense of life address the opening of *Science of Logic* section II.3.1 ("Life"), in which Derrida finds an exemplary formulation of the relationship between scientific and philosophical treatments of life. Hegel's passage begins as follows: "A comment may be in order here to differentiate the logical view of life from any other scientific view of it . . . to differentiate logical life as idea from natural life as treated in the *philosophy of nature*, and from life in so far as it is bound to *spirit*" (Hegel 2010, 677).[14]

In what follows, I draw attention to a page from the opening session of the 1964–1965 course on Heidegger in which Derrida develops a comparative reading of Heidegger's destruction against Hegel's concept of refutation. This page constitutes an initial inscription of the reading of the Hegelian system and of the concept of the philosophy of nature that Derrida further elaborates in the 1969–1970 course. Here he focuses on negativity as the minimal condition for productivity and thus for genesis and life in general. For Hegel, natural productivity is derived and secondary with respect to negativity, an image or metaphor of the latter.[15] Therefore, under the guise of negativity, the living logos constitutes the sense of the relations among regional discourses as well as the unfolding of the system that these relations make up. "And we know that this negativity," Derrida observes, "is essential to historical production, to the production of history in general, to production, to productivity in general" (Derrida 2016, 3). He quotes a text from Hegel's *Lectures on the History of Philosophy*, which includes the following passage:

> This refutation occurs in all development, hence also in the development of a tree from the seed. The blossom is

refutation of the leaves, such that they are not the highest or true existence of the tree. Finally, the blossom is refuted by the fruit. The fruit, which is the last stage, comprises the entire force of what went before. *In the case of natural things these levels occur separately* [Derrida's emphasis], because there nature exists in the form of division. In spirit too there is this succession, this refutation, yet all the previous steps remain *in unity*. The most recent philosophy, the philosophy of the current age, must therefore be the highest philosophy, containing all the earlier philosophical principles within itself. (2016, 3)

In the subsequent remarks, Derrida explains that natural productivity plays as an analogous of the spiritual one ("the natural example, the example of the tree, functions here only by analogy," 3), since the former reproduces a derived ("inferior," 3) feature of the latter. This inversion of appearances constitutes the sense of sciences—that is, philosophy. It is not by chance, I suggest, that Hegel's analogy resonates with the analogy between the generation of the Athenians from their fatherland and biological generation, as it is described in the Greek tradition of the *epitaphioi logoi* and referred by Socrates in Plato's *Menexenus*. The position of natural life—namely, its sense—is determined on the basis of the concept of negativity and spiritual productivity, understood as the conservation of past stages in the present. It is from this perspective that Hegel thinks of natural life as characterized by division—that is, by the fact that the beginning of a level (or a moment) coincides with the end of its antecedent (its cause or genitors) and, therefore, that this end is without return. Derrida's remarks on Hegel's text read:

> Nature is the form of division, and what is left behind or refuted, the seed for example, has simply expired [*périmé*], and is not *present* as such in the tree, the flower is not *present* in the fruit. In spirit, on the contrary, and philosophy is the highest form of spirit thinking itself, refutation is preserved in presence—what one can call by a term that is not Hegelian, but that does not, I believe, betray Hegel's intention, *sedimentation*—and the sedimentation of forces (Hegel talks here of forces) is a phenomenon not natural but spiritual. It is spirit itself. (2016, 3–4)[16]

It is worth observing that this concept of general productivity, as the conservation of the past stages in the present, reproduces Derrida's understanding of preformationism as the transference of the anthropomorphic concept of the *logos spermatikos* into the field of biology and the life sciences. In *Glas*, which I open up in the next chapter, Derrida directly addresses the implications that follow from taking such a concept of productivity, genesis and life in general, as the sense and development of the Hegelian system.

THE NEGATION OF CONSCIOUSNESS

This section explores Derrida's interpretation of the *Phenomenology*'s figure of the master as he elaborates it in the opening session of the 1969–1970 course, as well as in earlier texts, in light of his reading of Hegel's take on the equivocity of language. Derrida interprets mastery precisely as the Hegelian solution of the equivocity between natural and philosophical life—that is, as the process through which life, as the life of the concept, sublates natural life and thus constitutes the very process of this sublation. Ultimately, mastery designates the relationship between philosophy and the natural sciences and thus the position of the philosophy of nature within the system, its determination and sublation.

As explained by Hegel in the *Phenomenology of Spirit*, self-consciousness presents itself as such by showing that it is not attached to life and thus by staking its own life in a life-and-death struggle.[17] This struggle evolves into the relationship between the master and the slave when consciousness sublates its natural life—that is, preserves the latter in sublation (namely, *Aufhebung*), and thus survives. Hegel explains that "their act is an abstract negation [death, or the natural negation of life, which fails to achieve recognition], not the negation coming from consciousness, which supersedes in such a way as to preserve and maintain what is superseded, and consequently survives its own supersession [*Die Negation des Bewusstseins welches so aufhebt, dass es das Aufgehobene aufbewahrt und erhält und hiermit sein Aufgehobenwerden überlebt*]. In this experience," he concludes, "self-consciousness learns that life is as essential to it as pure self-consciousness" (Hegel 1977, 124–25). Derrida reads this passage as the very process of life, the sublation of natural life, the self-sublation of life, the life of the concept, the organization of regional discourses, and so forth. He remarks:

> The master risks losing natural life only with a view to retaining the life of the spirit in self-consciousness. It is in this operation that the unity or the passageway between the two meanings [*sens*] of the word *life*, which we evoked earlier on, is produced [*se produit*]. The recourse of the master to the other's work [*travail*] structures the very position of *Herrschaft* as well as the relation of mastery that philosophical discourse and philosophical self-consciousness entertain with the work of materially determined regions, especially with the most deadly [*mortifère*] one of these regions, the closest and most distant at once, mathematics. (Derrida 1969–1970, 1.18)

In a note dedicated to the aforementioned passage from the *Phenomenology of Spirit*, Derrida alludes to *Aufhebung* as to the process of sublation and auto-sublation that draws together two significations of life. "It is precisely this *natural-speculative concept-word* [*le concept-mot naturel-spéculatif*, my emphasis] of *Aufhebung*, which is self-contradictory and dialectical," he writes, "that puts into mutual relation these two lives [*vies*] in the movement of self-consciousness" (1.18). Therefore, the whole sense and organization of the Hegelian system is at stake in the survival of consciousness. "The Hegelian discourse . . . is inscribed again within the irreducible [*l'irréductible*] of a natural-speculative language, that is, in the play of its words, in a word-play" (1.19), Derrida concludes.[18]

This interpretation of mastery is also at work in earlier texts such as the 1964–1965 lecture course on Heidegger examined above and the well-known 1966 essay on Georges Bataille, entitled "From Restricted to General Economy: A Hegelianism without Reserve." The remarks developed in the Heidegger course revolve around the *Phenomenology*'s passage on the negation of consciousness (*Aufhebung*). Derrida explains the transition from abstract negation to *Aufhebung* as follows:

> In the first moment, consciousness, which has only the alternative choice of raising itself above or else saving its life, is placed before an *abstract* negation: in both cases one

loses, either as slave or as master, who in dying, also loses what he has won. So the master would have to keep what he loses (life), just as the slave, through labor, will also keep what he loses: freedom. To do so, he must pass from abstract negation to the *Aufhebung*: up to this point, says Hegel. (2016, 200)

Furthermore, the text points out that *Aufhebung* accounts for the sense of the relation between natural and philosophical life. "A passage from life to life, first of all," he notes, "from life in the sense of natural being-there, the life above which the point is to raise oneself through consciousness" (199). In the rereading of the Hegelian passage that Derrida offers in the aforementioned essay on Bataille, the movement that takes place within the equivocal term of life is designated as a surreptitious substitution of the natural and biological signification of life with the philosophical concept and signification of life. However, as Derrida suggests, it also constitutes the truth of life for Hegel (the life of the concept, etc.). The text reads:

Through a ruse of life, that is, of reason, life has thus stayed alive. Another concept of life had been surreptitiously put in its place, to remain there, never to be exceeded, any more than reason is ever exceeded (for, says *L'erotisme*, "by definition, the excess is outside reason"). This life is not natural life, the biological existence put at stake in lordship, but an essential life that is welded to the first one, holding it back, making it work for the constitution of self-consciousness, truth, and meaning. Such is the truth of life. (1978, 323)

As the text develops, Derrida further explores the concept of life as self-reproduction, as the interpretation of the genetic and zoological process of generation on the basis of the generation of consciousness, and, ultimately, as the *logos spermatikos*. This concept of life would be the element of the philosophical organization of regional discourses and thus the sense of the life sciences. "Through this recourse to the *Aufhebung*, which conserves the stakes, remains in control of the play, limiting it and elaborating it by giving it form and meaning (*Die Arbeit . . . bildet*)," Derrida explains, "this economy of life restricts itself to conservation, to circulation and self-reproduction as the reproduction of meaning" (1978, 323). As I highlight through my analysis of *Glas* in the next chapter, according to Derrida, it is on the equivocity of the term *Aufhebung* that the whole Hegelian system—that is, the latter's self-unfolding and the relation between philosophy and the life sciences—is grounded.

THE TWO DEATHS OF THE METAPHOR

In "White Mythology," Derrida demonstrates that, nowhere more explicitly than in Hegel, the system of the relations between nature and spirit, and thus between the natural sciences and philosophy, is entangled with the concept of the metaphor. To put it otherwise, nowhere more explicitly than in Hegel does metaphorization consist in the sense of natural life—that is, in the very solution of the homology between natural life and the life of the concept. In the chapter entitled "*Plus de métaphore*," Derrida formalizes the problem of the irreducible equivocity of philosophical language in terms of the tropic supplementarity of the metaphor—namely, of the impossibility of saturating the field of the metaphorical possibilities of philosophy without leaving outside of it "the metaphor without which the concept of the metaphor could not be constructed" (Derrida 1982, 219).[19] In particular, once more in the wake of Hegel, he draws attention to the equivocity of the founding concepts of philosophical language. "The difficulties we have just pointed out," Derrida remarks, "are accentuated with respect to the 'archaic' tropes which have given the determinations of a 'natural' language to the 'founding' concepts (*theoria, eidos, logos*, etc.)" (224). For instance, he takes up the concept of concept which "cannot not retain the gesture of mastery, taking-and-maintaining-in-the-present, comprehending and grasping the thing as an object" (224) and allows him to discuss the solution that Hegel has always already adopted for this case of equivocity. In fact, as he points out, this solution coincides with Hegel's understanding of philosophy as the sense of natural language and thus with the very organization and development of the Hegelian system. Derrida recalls that "noticing this fact, Hegel, in passing, defines our problem, or rather determines the problem with an answer indistinguishable from the proposition of his own speculative and dialectical logic" (225). I limit myself to quoting the final lines of the long passage from Hegel's *Aesthetics* examined by Derrida:

> In living languages the difference between actual metaphors (*wirklicher Metaphern*) and words already reduced by usage (*durch die Abnutzung*) to literal expressions (*eigentliche Ausdrücken, expressions propres*) is easily established; whereas in dead languages this is difficult because mere etymology cannot decide the matter in the last resort. The question does not depend on the first origin of a word or on linguistic development generally; on the contrary, the question above all is whether a word which looks entirely pictorial, deceptive, and illustrative has not already, in the life of the language, lost this its first sensuous meaning, and the memory of it, in the course of its use in a spiritual sense and been *relevé* (*AUFGEHOBEN HATTE*) into a spiritual meaning. (225)

What interests Derrida most is that, here, the process of *Aufhebung*, which he understands as the scheme of the organization as well as development of the Hegelian system, accounts for the solution of the equivocity of philosophical language. For this reason, he remarks that "*above all* [my emphasis] the movement of metaphorization (origin and then erasure of the metaphor, transition from the sensory meaning to the proper spiritual meaning by means of the figures) is nothing other than a movement of idealization" (226). It is understood in light of "the *master* [my emphasis] category of dialectical idealism" (226)—namely, *Aufhebung*, where, as we know, the term master (*catégorie maitresse*) alludes to a certain convergence between the figure of mastery and the relationship between philosophy, as dialectical idealism, and non-philosophical discourses. Metaphorization, understood as the solution of the problem of the equivocity of philosophical language and of its relationship with natural language, unfolds the philosophical claims of primacy and universality that, as Derrida points out in the 1969–1970 course, Aristotle ascribes to metaphysics. For this reason, Derrida's reading of the Hegelian remarks on the concept of concept ends as follows: "Nowhere is this system as explicit as it is in Hegel. It describes the space of the possibility of metaphysics, and the concept of metaphor thus defined belongs to it" (226).

A few pages later, Derrida focuses on the definition of the movement of metaphorization as the transition between two significations of the term "sense" and thus as the decision taken by Hegel on the problem of equivocity. Here he distances himself from Hegel's decision, which he recognizes, once again, as the most explicit accomplishment of the link between a certain philosophical tradition and the concept of the metaphor. He conjures up the figure of the enjoyment of thought before the natural-speculative word *Sinn* and opposes this figure to the interrogation of the unitary ground (that is, of the graphematic unity) that he had searched for since his *Introduction* to Husserl's *Origin of Geometry*. This ground is rather a text, the textuality of a text—what the *pharmakon* and *khōra* are for Plato's text—a weave of tropic movements and stories, syntax. Derrida writes:

> Already the opposition of meaning (the atemporal or non-spatial signified as meaning, as content) to its metaphorical signifier (an opposition that plays itself out within the element of meaning to which metaphor belongs in its entirety) is sedimented—another metaphor—by the entire history of philosophy. Without taking into account that the separation between sense (the signified) and the senses (sensory signifier) is enunciated by means of the same root (*sensus, Sinn*). One might admire, as does Hegel, the generousness of this stock, and interpret its secret *relève* speculatively, dialectically; but before utilizing a dialectical concept of metaphor, one must examine the double turn which opened metaphor and dialectics, permitting to be called *sense* that which should be foreign to the senses. (228)

Derrida brings to light a movement that has already been double and cannot be reduced to the transition within the word "sense" that accounts for the Hegelian solution, the sense of natural "sense," the *parti pris* of philosophy, etc. This movement is presupposed by the Hegelian solution, which wishes to erase it. It constitutes the very textuality of the Hegelian text. From the perspective of the anagrammatic writing formalized in "Plato's Pharmacy," the double turn mentioned in this passage accounts for the citational relations attached to the different functions of the same word—namely, with the *logoi*, stories, and significations that are inscribed within a unitary grapheme.[20] This turn also makes possible the story of the spiritual sense (that is, the sense of natural language) that philosophy tells about itself in order to escape its irreducible equivocity. The movements and turns highlighted by Derrida here can no longer be called metaphors precisely because they constitute the structure that metaphorization aims to dominate. They are graphical, anagrammatic, syntactical, and tropic. They are the textuality of the philosophical text and of its founding oppositions. Derrida describes them as follows:

> The constitution of the fundamental oppositions of the metaphorology (*physis/tekhnē, physis/nomos,* sensible/ intelligible, space/time, signifier/signified, etc.) has occurred by means of the history of a metaphorical language, or rather by means of "tropic" movements which, no longer capable of being called by a philosophical name—i.e. metaphors—nevertheless, and for the same reason, do not make up a "proper" language. It is from beyond the difference between the proper and the nonproper that the effects of propriety and nonpropriety have to be accounted for. By definition, thus, there is no properly philosophical category to qualify a certain number of tropes that have conditioned the so-called "fundamental," "structuring," "original" philosophical oppositions: they are so many "metaphors" that would constitute the rubrics of such a tropology, the words "turn" or "trope" or "metaphor" being no exception to the rule. To permit oneself to overlook this *vigil* of philosophy, one would have to posit that the sense aimed at through these figures is an essence rigorously independent of that which transports it, which is an already philosophical *thesis*, one might even say philosophy's *unique thesis*, the thesis which constitutes the concept of metaphor, the opposition of the proper and the nonproper, of essence and accident, of intuition and discourse, of thought and language, of the intelligible and the sensible. (228–29)

As we know, Derrida identifies the only thesis of philosophy with the position of an independent sense, which amounts to philosophy itself as the organization of regional discourses and thus the destruction of the syntactical layer of tropic movements. This sense is, above all, an incorruptible and self-reproducing life, since it has always already been the sense of natural life and of the life sciences. The aforementioned passage resonates with another text by Derrida, in the left-hand column of *Glas*, where he finds in the early dialectics of education the only thesis of the Hegelian philosophy—that is, the *being* dead of the father in the son's consciousness, and thus the thesis of incorruptible life, of the infinite germ, of the negation of consciousness, etc. In the next chapter, I offer an analysis of this moment in *Glas*.

In conclusion, I propose rereading the final pages of "White Mythology" from the perspective on the problem of homology that I have elaborated in this chapter. Derrida's essay ends by addressing the double fate of death prescribed to the metaphor. "Metaphor, then, always carries its death within itself" (271), he argues, thus having recourse to a formula that situates the double destruction that the metaphor has indefinitely been constructing for itself. In fact, this formula echoes a line from the last pages of the *Philosophy of Nature* of Hegel's *Encyclopedia* (section 375, "The Self-Induced Destruction of the Individual"), which reads as follows: "It is in this way that the animal brings about its own destruction" (Hegel 1970, 441). Here Hegel evokes the figure of the "inborn germ of death"—that is, the fate of death that the natural individual bears within itself—which, from a systematic perspective, secures the transition from the philosophy of nature and natural life to the philosophy of spirit and spiritual life. Postponing a deeper engagement with this text to the next chapter, I limit myself to remarking that, for Derrida, the germ of death, which he interprets as the natural and biological germ and, more generally, as the living individual *tout court*, is not only the element of Hegel's philosophy of nature but the general structure of genesis, from linguistics to biology, which I designated as the disseminated trace-seed. Therefore, like the mortal germ, the metaphor ties together two self-destructions that are related to two different perspectives of reading.[21] On the one hand, the death of the metaphor accounts for the death of natural life, which the sense and organization of the system, philosophy itself, bring about. On the other hand, the metaphor as well as natural life destroy themselves so long as they are detached from philosophical oppositions and thus generalized into a disseminated trace-seed.[22] Derrida has already announced this self-destruction a few pages prior, when he points out that tropic movements can no longer bear the name of metaphors. The key-features of these two self-destructions are described as follows:

(1)

> One of these courses [that the self-destruction of metaphor will always have been able to take] follows the line of a resistance to the dissemination of the metaphorical in a syntactics that somewhere, and initially, carries within itself an irreducible loss of meaning: this is the metaphysical *relève* of metaphor in the proper meaning of Being. The generalization of metaphor can signify this parousia. Metaphor then is included by metaphysics as that which must be carried off to a horizon or a proper ground, and which must finish by rediscovering the origin of its truth. (Derrida 1982, 268)

(2)

> The *other* self-destruction . . . passes through a supplement of syntactic resistance, through everything (for example in modern linguistics) that disrupts the opposition of the semantic and the syntactic, and especially the philosophical hierarchy that submits the latter to the former. This self-destruction still has the form of a generalization, but this time it is no longer a question of extending and confirming a philosopheme, but rather, of unfolding it without limit, and wresting its borders of propriety from it. (270)[23]

Therefore, not only does the first death affect the metaphor, as well as natural life, but it also concerns philosophy as the sense and organization of the system, it confirms and is induced by this organization. It is from this perspective, as Derrida remarks, that this death "is sometimes . . . death of a genre belonging to philosophy which is thought and summarized within it, recognizing and fulfilling itself within philosophy [for instance, of the philosophy of nature]" (271). But, for the same reason, the other death is inscribed on the back of the philosophical field of fundamental oppositions—namely, on the anagrammatic, syntactic and tropic layer of writing that philosophy wishes to master. "The death of a philosophy," so Derrida designates the other death, "which does not see itself die [as the life of the concept, the negating consciousness, the dead father, etc., do] and is no longer to be refound within philosophy [*et ne s'y retrouve plus*]" (271).

4

HEGELIANISM II
The Book of Life

> The absolute idea, as the rational concept that in its reality only rejoins itself, is by virtue of this immediacy of its objective identity, on the one hand, a turning back to life; on the other hand, it has equally sublated this form of its immediacy and harbors the most extreme opposition within.
>
> —Hegel, *Science of Logic*, 735

It is time to look into what Derrida calls the "book of life"—namely, the Hegelian text.[1] Let us start from a decisive note that he makes in the middle of *Glas*, on the left-hand column, a note on the progress of *Sittlichkeit* as it develops in Hegel's early works such as the *System of Ethical Life* and *Natural Law* (1802–1803). Derrida observes that, in Hegel's text, the process of spiritual life is illustrated through the recourse to images drawn from natural life. They are precisely "images"—that is, deforming imitations and metaphors—so long as nature is analogous to (and different from) spirit. Nature constitutes the initial moment of spiritual life (that is, of the spirit's return to itself), the general form of the spirit's other, and thus the spirit's being outside itself.[2] This analogy makes the recourse to images and metaphors possible and, more generally, regulates the being remarked of life throughout the ontological regions of the Hegelian system, as well as the metaphorical exchanges between these regions. In pointing to what remains unquestioned here,

Derrida formulates the eluded problems that I track in the following pages: Why is natural life an image and metaphor of spiritual life? Indeed, would it not be the opposite that holds? Why should we think of natural life on the basis of spiritual life?

> This whole process is described through what Hegel considers natural "images." He criticizes them less than explains their necessity: the regulated relation they maintain with their spiritual sense [*le rapport réglé qu'elles entretiennent avec leur sens spirituel*]. The animal and oriented figure of the Phoenix will be put back in its place by *Reason in History*. All the references to natural life and death imitate and deform the process of spiritual life or death. Everywhere the relation of nature to spirit is found: spirit is (outside itself) in nature; nature is spirit outside self. The finite metaphor, real organic life is impotent to receive all the spiritual divinity of *Sittlichkeit*; nevertheless it "already expresses in itself the absolute Idea, though deformed." It has within itself the absolute infinity, but "only as an imitative (*nachgeahmte*) negative independence—i.e., as freedom of the singular individual." (Derrida 1986, 103)[3]

The passage draws attention to a certain impotence (*Ohnmacht*) of natural life in relation to spiritual life. This impotence is inherent in the self-inequality of natural life, understood as the region of the originary separation, sexual contradiction, classification, and death.[4] Derrida raises the case of the phoenix. A few pages prior, he had observed that the phoenix describes the activity of the spirit in the element of *Sittlichkeit*. Hegel refers to the animal in an implicit fashion in order to account for "the ethical body" that "must incessantly repeat the spiritual act of its upsurge" (102).[5] However, he acknowledges only later, in *Reason in History* (1822–1830), that the phoenix is an image borrowed from natural life, a natural metaphor, and that this rhetorical operation hinges on the analogy between natural and spiritual life. Derrida recalls this moment as follows: "*Reason in History* specifies the limits it is advisable to recognize in the wingspan of the Phoenix": it is only "an 'image' of the spirit, an analogy drawn from the 'natural life' of the body . . . an 'oriental image'" (116–17). Derrida adds this passage as a note on the analogy that he suggests between the phoenix and the infinite germ of spiritual life, which, as it appears later, is incorruptible and self-inseminating. The *"like"* of the suggested analogy is in italics precisely as it accounts for a natural metaphor and a rhetorical operation. The metaphor of the germ comes back soon as it constitutes the *onto-theological* figure that is remarked throughout the ontological regions of the system and orients the play of correspondences among them, the metaphorical play *tout court*.

THE SYSTEMATIC FIGURE OF THE GERM

Derrida focuses on the natural image of the germ as it is conjured up in the account of "the determinations of the spirit" in *Reason in History*. Hegel explains that the essence of the spirit is activity, which consists in the self-equality of spiritual life—that is, in the spirit's reproducing itself, in its being at once the beginning and the end of this act of reproduction, father and son and, thus, family. This activity is not natural and is finite only in a restricted sense (namely, in man), just as the father-son relation and family are. Hegel writes:

> When the spirit strives (*strebt*) towards its own center, it strives to perfect (*vervollkommnen*) its own freedom; and this striving is fundamental to its nature. To say that spirit exists would at first seem to imply that it is a completed entity (*etwas Fertiges*). On the contrary, it is by nature active (*Tätiges*), and activity; it is its own product (*Produkt*), and is therefore its own beginning and its own end. Its freedom does not consist in static being (*ruhende Sein*), but in a constant negation of all that threatens to destroy freedom. The activity of spirit is to produce itself, to make itself its own object and to gain knowledge of itself; in this way it exists for itself. Natural things do not exist for themselves: for this reason, they are not free. (Derrida 1986, 24)[6]

Spiritual activity starts with the becoming-life of matter. Within this movement, the shift from the animal to human marks the transition from the animal moment of life to the spiritual one. To schematize this process, Hegel ascribes this transition to the fact that man inhibits the animal pressure (*Trieb* / *poussée*) by idealizing and interposing the ideal between pressure and satisfaction. He explains that "since he [man] knows the real (*Realen*) as the ideal (*Ideellen*) . . . this knowledge leads him to suppress (*hemmt*) his pressures; he places the ideal, the realm of thought, between the demands (*Drangen*) of the pressure and their satisfaction" (26).[7] Therefore, he concludes, by interrupting the animal auto-mobility in himself man breaks with natural life and frees freedom (what Derrida calls "the self-mobility of the spirit," 27).

At this point, Hegel retrieves the figure of the germ in order to account for the transition from natural to spiritual life and thus for the liberation of spiritual activity. As anticipated, Derrida observes that this figure is remarked throughout the regions of the system and that it relates one region to another on the basis of the analogy between nature and spirit (or, as Derrida calls it, speculative dialectics). He explains:

> The germ (*der Same*) is also, as germ, the onto-theological figure of the family. This concept (of) germ (*Same*, semen, seed, sperm, grain) regularly enters on the scene in speculative dialectics, *in places and regions of the encyclopedic discourse that are not at once homologous and distinct* [my emphasis], whether of the vegetal, biological, anthropological, or the onto-logical order in general. Among all these orders, speculative dialectics assures a system of figurative correspondences. (Derrida 1986, 27–28)

The germ describes the essence of spirit as activity and thus consists in the element of the metaphorical exchange between the regions of the system. Derrida notes that "the figure of the seed is immediately determined: (1) as the best representation of the spirit's relation to self; (2) as the circular path of a return to self" (28). Indeed, the germ has already been at stake in Hegel's previous determination of the essence of the spirit as a self-identical and self-reproducing life, as the father-son relation and family.[8] In the text commented by Derrida, Hegel borrows the image of the germ from nature in order to illustrate the activity of the spirit. What makes this borrowing possible is the analogy between natural and spiritual life. In the Hegelian philosophies of nature that are examined later, the germ marks the end of an individual and the beginning of a new one and thus the positive self-relation of natural life within the limits of its self-inequality. Therefore, here the germ occurs as a deforming imitation and a metaphor of spiritual life. Hegel's text reads:

> Only the returned-home-to-self is subject, real actuality. Spirit exists only as its own result. The example of the seed (*die Vorstellung des Samens dienen*) may help to illustrate this point [*A titre d'éclaircissement (ou d'illustration:* zur Erläuterung*), on peut se servir de la représentation de la semence (*die Vorstellung des Samens dienen*)*]. The plant begins with the seed, but the seed is also the result of the plant's entire life, for it develops only in order to produce (*hervorzubringen*) the seed. We can see from this how impotent life is (*die Ohnmacht des Lebens*), for the seed is both the origin [*commencement*] and the result of the individual; as the starting point and the end result, it is different and yet the same, the product of one individual and the beginning of another. Its two sides fall asunder like the simple form (*Form*) within the grain (of wheat: *Korn*) and the whole course of the plant's development [*comme la forme* (Form) *de la simplicité dans le grain (de blé:* Korn*) se sépare au cours du développement de la plante*]. (28)

The impotence of natural life comes back in this passage. It presupposes the analogy with spiritual life according to which the natural germ is an image, a rhetorical operation in the text. Hegel explains the shift from the animal to human in light of the spiritual activity illustrated by the metaphor of the germ. While the animal belongs to nature and thus undergoes a natural development whose telos is a death without return, the human is already spirit and thus reproduces itself. In other words, they belong to different regions of the system that are tied together by the systematic figure of the germ and according to the nature-spirit analogy. Hegel observes: "Its growth (*Wachstum*) [of the animal] is a merely quantitative increase in strength (*Erstarken*). Man, on the other hand, must make himself what he should be; he must first acquire everything for himself, precisely because he is spirit" (28). In natural life, the germ neither reproduces nor returns to itself; it is not self-identical (that is, father and son at once, family). Therefore, Derrida remarks, "there is no natural family, no father/son relation in nature" (29).[9]

Derrida draws the same conclusion from the reading of the philosophy of nature of the *Encyclopedia*. Hegel conceives of the relation between natural and spiritual life according to the quasi-remark of *copulation* in the sexual reproduction of animals as well as in the syllogism. The figure regulates the correspondence between two ontological regions. The process of copulation is analogous to the copula of syllogism as it amounts to the self-equality of the genus—that is, of the universality in the animal—and yet, it differs from copula itself since copulation brings about the death without return of individuals. Spiritual life begins with the first moment of *Sittlichkeit*, the human family.

> The *Encyclopedia* states it precisely: in the animal kingdom, generation, the sex relationship, the process of copulation that, like a syllogism's copula, gathers together the genus with itself—they all engulf individuals in a death straight out [*sans phrase*]. Unlike the human, rational family, animal copulation does not give rise to any higher determination. Animal copulation leaves behind itself no monument, no burial place, no institution, no law that opens and assures any history. It names nothing. (Derrida 1986, 12)[10]

Returning to the progress of the spirit in *Reason in History*, Derrida concludes that the human is already spirit so long as it reproduces itself and thus is the result of its own activity. Therefore, the human germ is already a spiritual one, self-identical life, father-son relation, and family. It is the onto-theological figure of the system and accounts for the latter's organization and unfolding. As Derrida puts it, "It is, more than the plant or animal, its own proper product, its own son, the son of its work . . . the human individual is descended from its own germ. It conceives itself" (29).

In the wake of Hegel's text, Derrida explains that the human individual and, more generally, the human and rational family are, in turn, examples and finite images of the properly called spiritual life, "of the infinite father-son relation, of the relation of infinite spirit freely relating to itself" (29). The most sublime example of this relation—or, better, the truth from which all examples derive—is provided by the Christian God. "In the first place," Hegel writes, "he is Father"; "secondly, he is . . . a dividing himself into two [*ein sich Entzweiendes*], the Son." "But this other than himself," Hegel continues, "is equally himself immediately; he knows himself and intuits himself in that—and it is this self-knowledge and self-intuition which constitutes the third element, the Spirit itself" (30–31). Therefore, the spirit is designated properly by a self-identical life and an infinite germ: it constitutes the sense of metaphorical exchanges and is remarked throughout the regions of the system. It is the Father-Son relation, Derrida explains, the element in which the seed returns to the father.[11] Recalling the ontological figures of germ and copulation, Derrida describes this relation as a process of self-reproduction, which implies neither the death without return of the individual nor sexual difference, and thus as "self-fellatio," "self-insemination" or "self-conception" (31). A few pages later, commenting on Hegel's reading of John's gospel in the early *Spirit of Christianity and its Fate* (1799), Derrida points out that the spirit, understood as the Father-Son relation and the element of the self-identical and self-reproducing germ, is the name. Hegel translates the name proper of a man (*onoma*), which makes man recognize itself as the son of God, by relation (*Beziehung*).[12] Derrida justifies this operation by observing precisely that "the name, the relation, the spirit (Hegel sometimes translates *onoma* by *spirit*) is the structure of what returns to the father" (79).

In the passage from *Reason in History* dedicated to man, Hegel identifies the human example of the Christian family as a process of education/formation (*Bildung*). Derrida investigates this articulation of human family and education through a close reading of the Hegelian account of the third *Potenz* ("Possession and Family") in the early *Philosophy of Spirit 1803–1804*. Hegel explains that the speculative dialectic of the wedding consists in the formation of the consciousness of the son—that is, in the process of education (*Erziehung*):

> It is in the child that the partners recognize themselves as one, as being in one consciousness, and precisely therein as *superseded* [my emphasis], and they intuit in the child their own coming supersession. . . . As they *educate* it, they posit their achieved consciousness in it, and they generate their death, as they bring their achievement to living consciousness. (Hegel 1979, 323)

Education is the positing and negation (*Aufhebung*) of the consciousness of genitors in the living consciousness of the son. In analyzing this passage, Derrida measures education against animal copulation, which brings about the genitors' death without return and no further determination. Education marks the human and rational moment of family, of the father-son relation and thus of the spirit's self-insemination. It is in light of this play of correspondences between the animal and the rational that Derrida observes that "the natural child does not bear [*ne porte pas*] the death of its genitors. So the death of the parents *forms* the child's consciousness" (Derrida 1986, 132). This bearing must have to do with the *Aufhebung* of the parents and, more precisely, of the father.[13] Therefore, the element of the spirit has already taken place: a father who returns in the son, an incorruptible and self-reproducing seed, spiritual life. Unfolding the analogy with natural life, Derrida explains that "the relieving education interiorizes [idealizes] the father. Death being a relief, the parents, far from *losing or disseminating themselves without return* [my emphasis], 'contemplate in the child's becoming their own relief'" (133).[14] The spiritual seed is not disseminated and thus it returns to itself. It is not a metaphor but the onto-theological figure that is remarked in natural life and on the basis of which the natural germ is determined. The spiritual seed, as the *Aufhebung* of the death of the father, as its *being* dead, is the thesis (the presupposition, the *parti pris*, etc.) of Hegelian philosophy. Derrida writes:

> Ideality is death, to be sure, but to be dead—this is the whole question of dissemination—is that *to be* dead or to be *dead*? . . . if death is the being of what is no more, the no-more being, death is nothing, in any case is no longer death. Its own proper death, when contemplated in the child, is the death that is denied, the death that *is*, that is to say, denied. When one says "death is," one says "death is denied," death is not insofar as one *posits* it. Such is the Hegelian *thesis*: philosophy, death's positing, its pose [*la philosophie, la pose de la mort*]. (1986, 133)[15]

At the end of this analysis, I suggest that another dissemination—that is, another circulation of the singular germ in the element of natural and biological life—is uncovered when the natural germ and, more generally, natural life are no longer determined out of the analogy between nature and spirit.

THE TREE OF LIFE

The reading of the metaphorical exchange at play in the Hegelian system culminates in the analysis of the remark of the vegetal image in *The Spirit of Christianity*. A tree is marked at least three times in this text, in order to say the truth about the father-son relation, the bond between God and Jesus Christ. Is it a natural image? Is it a metaphor? Is it not, rather, what allows us to think and speak about the bond—that is, about the element of self-return and thus about the activity of the spirit? An onto-theological figure? What is the relation between this tree and the natural tree? These questions underlie Derrida's analysis.

Derrida focuses on the fourth section of the *Spirit of Christianity*, which follows Hegel's reading of the Last Supper and is dedicated to "The Religious Teaching of Christ." The first tree of life is quoted from a note on the text (that is not included in the English edition of the *Spirit of Christianity*):

> This relation of a man to God in which is found the son of God, similar to the relation of branches, of foliage and fruits to the trunk their father, had to rouse the deepest indignation of the Jews, who had maintained an insurmountable abyss between human being and divine being and had accorded to our nature no participation in the divine. (Derrida 1986, 73)

This figure of the tree illustrates the bond between God and Jesus Christ—that is, life itself, life as the return of the whole in the part. Derrida explains: "The bond (*Band*) [what Hegel determines as life] holds God and Jesus together, the infinite and the finite; of this Jesus is a part, a member (*Glied*), but a member in which the infinite whole is integrally regrouped, remembered [*se remembre, se rappelle intégralement*]" (72).[16] In the wake of Hegel's passage, Derrida suggests that the tree of life says what remains inaccessible to the Jews, the self-circulation of the whole from the seed throughout its parts.[17] Therefore, this tree is neither a natural image nor a metaphor, as the natural seed and the phoenix are. If we take up the sections of the organics dedicated to the plant in the philosophy of nature (sections 343–349), we verify the impotence of vegetal life with respect to the self-identical and self-reproducing life of the spirit and thus to the God-Jesus Christ relation. The tree of life is the onto-theological figure that is presupposed by the metaphorical determinations distributed throughout the system, by the metaphorical play *tout court*. In having recourse to the figure of the tree, Hegel does not speak about spiritual life through a biological discourse. Rather, this figure says what makes any regional discourse possible, the element of metaphorical exchanges, and the sense of natural life. Therefore, Derrida suggests that, rather than a metaphor, the tree of life speaks about (spiritual) life as metaphoricity—that is, as a certain bond between the whole and the part. "When one feels it from inside [what the Jews could not do]," he writes, "one knows that life is metaphoricity, the alive and infinite bond of the whole thought in its parts [*lien vivant et infini du tout pensé dans ses parties*]" (73). Here, the Jews take the place that corresponds to their understanding of life; in other words, they are determined according to the analogy that regulates the exchange between the regions of the system and thus according to the onto-theological and systematic "metaphor" of (spiritual) life. At the beginning of the fourth section of the *Spirit of Christianity*, Hegel notes that "to the Jewish idea of God as their Lord and Governor, Jesus opposes a relationship of God to men like that of a father to his children" (73).[18]

A few pages later, Hegel conjures up the tree of life, once more, in order to reaffirm the idea of life as the bond of the whole and the part. The tree divides into parts that are themselves the whole, as well as God divides (*Entzweiung*) into the other in which he recalls himself. "Life is a strange division producing wholes" (77), Derrida remarks. The Hegelian text reads:

> . . . a branch of the infinite tree of life (*ein Zweig des unendlichen Lebensbaumes*). Each part, to which the whole is external [*en dehors de laquelle est le tout*], is at the same time (*zugleich*) a whole, a life. (Derrida 1986, 77)[19]

Again, Derrida seems to suggest that this is not merely a metaphor, but it describes the movement of life and the organization of the system. "Here the 'metaphor' of the tree," he observes, "turns up again as a family metaphor: the genealogical tree in a radical sense" (77). It is likely that the tree of life is the most sublime example and, more properly, the truth of life, and that the other examples derive from this tree and are organized in relation to it. The tree of life makes us think and speak about the analogy (life as metaphoricity and whole-part reconciliation) out of which these examples are organized and related within the system. Derrida draws attention to the Hegelian *zugleich*, which literally means "at the same time" (*en même temps*)—"the structural at once (*simul*) of the living whole and morsel" (78)—because it keeps the secret of life as metaphoricity (the secret that the Jews cannot understand).

Before evoking the third tree of life, Hegel recalls the distinction between the Jews, who are situated in the region of the concept and thus remain on this side of life, and the Christians, who are in the region of the whole-part relation (namely, metaphoricity)—the system itself—and thus look at life from inside. He explains:

> The relation of a son to his father is not a unity, a concept (as, for instance, unity or harmony of disposition, equality of principles, etc.) . . . a unity which is only a unity in thought and is abstracted from life. On the contrary, it is a living relation of living beings, a likeness of life [*gleiches Leben*, which Derrida translates by *l'egalité de la vie*]; simply modifications of the same life, not the opposition of essences, not a plurality of absolute substantialities [*mais une pluralité de substantialités absolues*]. Thus the son of God is the same essence [*le même être*] (*Wesen*) as the father. (Derrida 1986, 80)[20]

For the third time, Derrida suggests that the metaphoricity of the tree is strictly required. "Since this unity [that of Father-Son] cannot be stated in the understanding's abstract language," he writes, "it requires a kind of metaphoricity" (80). Once again, acknowledging the necessity of the figure of the tree does not mean that Hegel merely speaks about spiritual life through a natural image—that is, through natural life. Rather, echoing Hegel's second preface to the *Science of Logic*, we may suppose that thought is delighted of finding the traces of speculative spirit in natural life. As Derrida points out, this time the tree of life consists of the self-circulation of the whole in all parts, as well as of the equality and thus permutability of one part with another. Hegel describes the tree as follows:

> A tree which has three branches makes up *one single* tree (einen *Baum*); but every son of the tree, every branch (and also its other children, leaves and blossoms) is itself a tree. The fibers bringing sap to the branch from the trunk are of the same nature (*gleichen Nature*) as the roots. If a tree is set in the ground upside down it will put forth leaves out of the roots in the air, and the boughs (*Zweig*) will root themselves in the ground. And it is just as true to say that there is only *one single* tree here as to say that there are three. (81)[21]

Derrida draws attention to the implications that follow from the permutability of parts. Translated into the father-son relation, this permutability entails that the son always becomes "the father of the father" as well as the father becomes "the son's son" (81). Before discussing this point, I propose reconsidering the tree of life and, more generally, the vegetal metaphor of (spiritual) life in relation to the system and to how Hegel wants it to be read. If the tree of life speaks about spiritual life and thus about the living organization of the system, what would the consequences be for our reading of Hegel's text? In other words, does the tree of life say something about how, according to Hegel, we should read Hegel?

Finally, Derrida analyzes the metaphor of the tree of life from the perspective of rhetoric. It can be divided into two marks of life: a "semantic tenor," corresponding to the "life of spirit," and a "metaphorical vehicle," which amounts to natural life. "The life of the spirit," he explains, "is named through natural life in which it grows [*végète*]" (82), where "vegetating" accounts for the spirit's being outside itself. What secures this metaphor, this onto-theological figure, and thus the articulation of the two marks, is the nature-spirit analogy—namely, the presupposition that nature is the spirit's being outside itself and that the spirit returns to itself through the negation of nature. "Between the two lives, analogy makes metaphor possible" (82), Derrida notes. Therefore, "this double mark of life describes the structure of all life, the living organization of the Hegelian system" (82).

The Hegelian text—namely, the system—must reproduce this metaphor and double mark. Derrida gives an index of this necessity by referring to the *Science of Logic*, which has a privileged place within the system. It is not by chance that, in the last section, dedicated to

"The Idea," the reader finds the mark of life at the beginning as well as at the end of the syllogism, that is, as the first moment (natural and immediate life) and its movement (true and absolute life). Derrida suggests interpreting this double mark (or remark) as the spirit-nature relation and thus as the very metaphor of life:

> In this syllogism of the Idea, life first appears as a natural and immediate determination: the spirit outside self, lost in naturality, in natural life that itself constitutes a "smaller" syllogism [the French has no "in": *l'esprit hors de soi, perdu dans la naturalité, vie naturelle* . . .]. The immediate Idea has the form of life. But the absolute Idea in its infinite truth is still determined as Life, true life, absolute life, life without death, imperishable life, the life of truth. (82)

The "proper, literal sense" of life, Derrida explains, resides neither in natural life nor in the absolute one, to the extent that "life produces itself as the circle of its reappropriation, the self-return before which there is no proper sense" (82). Therefore, the metaphor of life is more than a simple metaphor as we cannot render the circle and self-return otherwise. If we hold on to the hypothesis that the system unfolds itself into a tree of life, we discover an internal prescription. "The Hegelian system," Derrida writes, "commands that it be read as *a book of life* [my emphasis]" (83).[22] From this perspective, Derrida addresses the case of Bernard Bourgeois's book on Hegel at Frankfurt. This book takes up the categories of the biological paradigm of preformationism by which the young Hegel would be the preconfiguration of the adult one, in which he finds his accomplishment.[23] However, although very Hegelian, this is not Hegelian enough, because it looks at the system from the outside, from a Jewish point of view. It produces divisions that do not reconcile within the whole and are neither equal nor permutable as they are in the tree of life. Bourgeois's reading does not admit that the tree is turned upside down, that leaves begin to root, and roots begin to flower. Derrida concludes:

> Nothing more Hegelian. But nothing less Hegelian: in distinguishing the old from the young, one sometimes dissembles the systematic chains of the "first" texts; and above all one applies a dissociating and formal analysis, the

viewpoint of the understanding in a narration that risks missing the living unity of the discourse; how does one distinguish philosophically a before from an after, if the circularity of the movement makes the beginning the end of the end? And reciprocally? The Hegelian tree is also turned over; the old Hegel is the young Hegel's father only in order to have been his son, his great-grandson. *The risk, then, is the Jewish reading* [my emphasis]. (84)[24]

In a lateral addition to the text, Derrida further interrogates this permutability. He observes that it brings about the ruin of the system insofar as it destroys the determinations and oppositions that secure the system's organization and development. The tree of life, with the permutability of its features, is a metaphor that cannot satisfy and thus escapes us: it is not the mere self-return of sense. Perhaps, as Derrida seems to suggest ironically, we should leave metaphors such as the tree of life to theologians.[25] The passage reads:

> The pure activity of the spirit—the spirit produces itself—a little farther on induces the spirit's assimilation to the father who produces or gives himself, by doubling himself, a son. I am my father my son and myself. My name is my father [*Je m'appelle mon père*]. But the giving-producing-doubling-himself insinuates into the pure activity an inner division, a passivity, an affect that obscurely breaches/broaches the father's paternity and begins to ruin all the determinations and oppositions that form system with it. All the family significations then set about passing (away) in each other, and nothing can stop them. Such is the play of spirit with itself, as soon as it begins to stretch, strain itself. (24)

THE CIRCULATION OF SINGULAR GERMS

The analysis of the last sections of Hegel's philosophy of nature is introduced by a remark on the position of *Naturphilosophie* within the organization of the system. Derrida writes that "the philosophy of nature is the system of this fall [of the spirit] and of this dissociation into exteriority. The philosophy of spirit is the system of the relief [*relève*] of the idea that calls and thinks itself in the ideal element of universality" (108). The last sections of the philosophy of nature secure the transition from one region of the system to another and the metaphorical exchange between the two. As Derrida points out, in the Jena philosophy of nature (JPN, 1803–1804, 1805–1806), as well as in the *Encyclopedia* philosophy of nature (EPN), the transition from natural to spiritual life is carried out through disease and death, which put an end to the self-inequality of natural life. In the last sections of JPN, Hegel identifies disease as the "dissolution of the whole" (*Auflösung des Ganzen*; Hegel 1987, 168), a "negative force" (*negative Kraft*, 168), and a "critical dissociation" (*kritische Ausscheiden*, 167) that bear the spirit's becoming. Referring to this passage, Derrida observes that the spirit itself—and thus the very becoming of life and the system—operates within natural and biological life as the work of negativity and dissociation:

> In the dissociation of the natural organization, the spirit reveals itself. It was working biological life, like nature in general, from its negativity and manifests itself therein as such at the end; spirit will always have been nature's essence; nature is within spirit as its being-outside-self. In freeing itself from the natural limits that were imprisoning it, the spirit returns to itself but without ever having left itself. A procession of returning (home). (Derrida 1986, 109)[26]

The same negativity is at work in the last sections of EPN so that natural life and the life of the spirit take their places as the Hegelian system develops. "This joint [*conjuncture*] will assure, in the circle of the Encyclopedia, the circle itself, the return to the philosophy of spirit" (109), Derrida notes. In this circle he finds the accomplishment of teleology—namely, of the concept of internal finality that Aristotle discovered in nature (*physis*) and Hegel reformulates as an unconscious instinct (EPN section 360). At this point, the reading of the last sections of the organics, which are grouped under the heading of "The Process of the Genus" (sections 367–75), begins.

Paraphrasing the opening section of this part of the organics, Derrida draws attention to the process by which the simple universality of the genus, which is constituted by an originary division, strives to return to itself through the negation of universality itself and that of the natural living (namely, its death). "Genus," he writes, "designates the simple unity that remains (close) by itself in each singular subject," but "as it is produced in judgment, in the primordial separation (*Urteil*), it tends to go out of itself in order to escape morseling, division, and to find, meet itself, again back home, as subjective universality" (109).[27] In the addition to this section, Hegel points out that, since "the genus is related to the individual in a variety of ways," the genus-process takes on different forms or processes (division into species; sex-relation; disease and natural death) that amount to "the different ways in which the living creature meets its death" (Hegel 1970, 411). In the first case, the genus is divided into species that differentiate through their reciprocal negation. As Derrida observes, in this case "the genus produces itself through its violent auto-destruction" (Derrida 1986, 109).[28]

However, the genus is not only a hostile relation between singularities but also the positive relation of singularity to itself, which is called sex-relation. As Hegel puts it, "The genus is also an essentially affirmative relation of the singularity to itself in it; so that while the latter, as an individual, excludes another individual, it continues itself in this other and in this other feels its own self" (Hegel 1970, 411).[29] Sex-relation consists of the following moments: (a) it begins with a "need" and a "feeling of lack" because of the originary and constitutive separation between the singular individual and the genus immanent in it; (b) the genus operates in the singular individual as a tension to resolve its inadequacy and thus as a striving to integrate itself with the genus, to close the genus with itself and bring it into existence; (c) the accomplishment of this operation is copulation (*Begattung*).[30] In the wake of Hegel's account, Derrida anticipates the result of sex-relation—that is, the originary separation and the constitutive self-inequality of the genus. "In the same stroke," he observes, "pressure [*la poussée*] tends to accomplish just what it strictly reduces, the gap of the individual to the genus, of genus to itself in the individual, the *Urteil*, the primordial division and judgment" (Derrida 1986, 110).

A long addition follows the section of EPN dedicated to sex-relation. "In it Hegel treats of the sexual difference" (110), Derrida points out. Hegel explains that the union of sexes—namely, copulation—consists in the development of the simple universality that is implicit in them, the genus itself, and thus in the highest degree of universality that an animal can feel. However, neither this universality becomes the object of a theoretical intuition, such as thought or consciousness, nor the animal reaches the free existence of the spirit. The addition reads:

> The process consists in this, that they become in reality what they are in themselves, namely, one genus, the same subjective vitality. Here the Idea of Nature is actual in the male and female couple; their identity and their being-for-self, which up till now were only for us in our reflection, are now, in the infinite reflection into self of the two sexes felt by themselves. This feeling of universality is the highest to which the animal can attain; but its concrete universality never becomes for it a theoretical object of intuition: else it would be Thought, Consciousness, in which alone the genus attains a free existence. The contradiction is therefore, that the universality of the genus, the identity of individuals is distinct from their particular individuality; the individual is only one of two, and does not exist as unity but only as a singular. The activity of the animal is to sublate this difference. (Hegel 1970, 412)

Derrida focuses on the contradiction highlighted by Hegel at the end of the passage. This contradiction has to do with sexual difference insofar as the latter divides the universality of the genus, which is the same as the identity of the individual, from particular individuality, which is one of the two or a (sexually differentiated) singularity. "Sexual difference," Derrida remarks, "opposes unity to singularity and thereby introduces contradiction into genus or into the process of *Urteil*" (Derrida 1986, 111). Therefore, sex-relation is the *Aufhebung* of sexual difference and of the inherent contradiction to the extent that it resolves as well as conserves them (and, more generally, natural life). Here, Derrida finds the knot of the equivocity between natural and spiritual life, the point where the whole structure of the Hegelian system—the presupposition of spiritual life and its being remarked throughout the regions of the system—trembles. He observes that "*Aufhebung* is very precisely the relation of copulation and the sexual difference" and, a few sentences later, that "the *Aufhebung* of the sexual difference is, manifests, expresses, *stricto sensu*, the *Aufhebung* itself and in general" (111). In the following section dedicated to sex-relation, Hegel explains that the "product" of copulation (as "the negative identity of the differentiated individuals") is the "realized genus" and thus "an asexual life" (Hegel 1970, 414)—namely, the *Aufhebung* of the contradiction of sexual difference. However, this occurs only in principle, he adds, because "the product . . . is itself an immediate singular, destined to develop into the same natural individuality, into the same difference *and* perishable existence" (414; I emphasize the "and" as it accounts for the irreducible articulation of sexual contradiction and biological life). Hence, the generic process develops through a "spurious infinite process" without being liberated from its primordial and constitutive separation (and contradiction). In the addition to the section, Hegel demarcates the genus from the spirit, which "preserves itself" and "exists in and for itself in its eternity" (414). Therefore, the relation between biological and spiritual life is understood on the basis of the analogy between the natural, singular, and mortal germ of a species and the incorruptible and infinite germ of God.

After the sections dedicated to the division of the genus into species and to sex-relation, Hegel proceeds to analyze the third form of the genus-process and thus the third way in which the animal dies, so that natural life and the spiritual one find their proper place and the circle of encyclopedia is secured. "Another negativity works (over) the indefinite reproduction of the genus, the nonhistoricity and the faulty infinite of natural life" (Derrida 1986, 115), Derrida notes. "The genus observes itself only through the decline and the death of individuals: old age, disease and spontaneous death" (115).[31] In the following pages, I focus on Derrida's reading of the last sections of EPN, which are dedicated to the natural death of natural life. Hegel's argument is that, apart from the inequalities caused by disease, there is a self-inequality of natural life that the animal cannot overcome—the opposition between implicit universality and natural singularity: that is, the contradiction within the originary separation—and thus the genus-process operates as a negativity within the animal itself, as the spirit's becoming.

> The animal, in overcoming and ridding itself of particular inadequacies, does not put an end to the general inadequacy which is inherent in it, namely, that its Idea is only the immediate Idea, that, as animal, it stands within Nature, and its subjectivity is only implicitly the Notion but is not for its own self the Notion. The inner universality therefore remains

opposed to the natural singularity of the living being as the negative power from which the animal suffers violence and perishes, because natural existence (*Dasein*) as such does not itself contain this universality and is not therefore the reality which corresponds to it. (Hegel 1970, 440)

As Hegel explains in the addition, the "necessity of death" does not depend on a particular cause but on the "necessity of the transition of individuality into universality," which constitutes the genus-process (441). Therefore, within natural and biological life, the genus-process acts as spirit. Following the Hegelian argument, Derrida concludes that there is a natural death of natural life and he calls it classification. "There is natural death," he writes, "it is inevitable for natural life, since it produces itself in finite individual totalities. These totalities are inadequate to the universal genus and they die from this. Death is this *inadequation* [*inadéquation*, my emphasis] of the individual to generality: death is the *classification* itself, life's inequality to (it)self [*l'inégalité à soi de la vie*]" (Derrida 1986, 116).

In the subsequent section, which is dedicated to "The Self-Induced Destruction of the Individual," Hegel identifies the inequality immanent in the animal (the difference between its singularity and the implicit universality of the genus) as a constitutive inequality and thus as a prescription of death: "The disparity between its finitude and universality is its original disease and the inborn germ of death, and the removal of this disparity is itself the accomplishment of this destiny" (Hegel 1970, 441). As Derrida points out, this constitutive inequality bears within itself sexual difference and the contradiction inherent in it. Therefore, the determinations of originary disease and mortal germ should also be extended to them. They "inhabit the same space" (Derrida 1986, 116), Derrida observes, the space of life's self-inequality, of natural and biological life, which Hegel understands as an image or a metaphor of spiritual life. Furthermore, if the germ is literally situated in this space, then, Derrida suggests, the Hegelian "germ of death" constitutes "a tautological expression" (116). Naming the individual singular that is inadequate, sexually differentiated, classified, and so forth, the germ is always a germ of death. Dissemination is the circulation of singular and mortal germs in the space of natural and biological life. Derrida explains:

> At the bottom of the germ, such as it circulates in the gap [*écart*] of the sexual difference, that is, as the finite germ, death is prescribed, as germ in the germ [*en germe dans le germe*]. An infinite germ, spirit or God engendering or inseminating itself naturally, does not tolerate sexual difference. Spirit-germ disseminates itself only by feint. In this feint, it is immortal. *Like* a phoenix. (1986, 116–17)

The passage recalls the ontological figure of the germ on the basis of which Hegel conceives of the relation between natural and spiritual life as well as of the living organization of the system. As we know, not only is the spiritual germ neither singular (separated from the genus, sexually differentiated, classified, disseminated) nor mortal, but it is also the *Aufhebung* of the natural germ, the spirit's return to itself through the death of the natural germ. Therefore, it describes the double mark of life as well as the circle of the *Encyclopedia*.[32] Derrida notes that it "disseminates itself only by feint." He thus rewrites an expression from Feuerbach's *Towards a Critique of Hegel's Philosophy*, in which Hegel's idea of alienation is described as follows: "The estrangement (*Entäußerung*) of the idea is, so to speak, only a feint [*une feinte*]; it makes believe, but it produces not in earnest; it is playing" (Derrida 1981, 40).[33] At this point, the question about the double mark of life comes back: is the spiritual and infinite germ a natural image (or a metaphor) of spiritual life? Where does the proper sense of germ (and of life) lie? Does not metaphor precisely describe the self-return of the spiritual germ through the *Aufhebung* of the natural one? As Derrida puts it:

> Then, the germ, finite germ of sexual difference, the germ of death, is it a metaphor of the infinite germ? Or the contrary? The value of the metaphor would be impotent to decide this if the value of the metaphor was not itself reconstructed from this question [*depuis cette question*]. (Derrida 1986, 117)[34]

A NOTE ON CLASSIFICATION

I conclude the analysis developed in the previous section by speculating on the term "classification" that Derrida emphasizes above. Hegel treats classification in the "Remark" and in the addition that follow the section on the particularization of the genus into species. He understands it as the main concern of zoology, which searches for "sure and simple signs of classes, orders, etc., of animals [and thus, as he explains a sentence later, of "artificial systems"] for the purpose of a subjective recognition of them" (Hegel 1970, 423). However, classification is a difficult task, Hegel explains, to the extent that nature is the self-externality of the idea, and thus that the existence of the idea in nature is determined by manifold conditions and circumstances and can take on the most inadequate form.[35] In the examined passage from *Glas*, classification accounts for the self-inequality of natural and biological life and, therefore, for the constitutive process of *Urteil*, sexual contradiction and dissemination in general. I highlight the resonance between this occurrence in *Glas* and the concept of classification that Derrida elaborates in a specific moment of *Grammatology*. Part 2 of this book begins with the analysis of Levi-Strauss's observations on the battle of proper names among the population of the Nambikwara. Here Derrida describes classification as the necessary inscription of the proper (or family) name and thus as the trace-seed—that is, the minimal condition for genesis in general. Furthermore, in a passage where he holds on to a Hegelian lexicon, Derrida unfolds this condition as the other, the outside, and thus the nature from which the life of the spirit can neither liberate nor return to itself. The passage reads:

> Thus the name, especially the so-called proper name, is always caught in a chain or a system of differences. It becomes an appellation only to the extent that it may inscribe itself within a figuration. Whether it be linked by its origin to the representations of things in space or whether it remains caught in a system of phonic differences or social classifications apparently released from ordinary space, the proper-ness of the name does not escape spacing. Metaphor shapes and undermines the proper name. The literal (*propre*) meaning does not exist, its "appearance" is a necessary function and must be analyzed as such in the system of differences and metaphors. The absolute parousia of the literal meaning [*sens propre*], as the presence to the self of the logos within its voice, in the absolute hearing-itself-speak, should be *situated* as a function responding to an indestructible but

> relative necessity, within a system that encompasses it. That amounts to *situating* the metaphysics or the onto-theology of the logos. (Derrida 1974, 89)[36]

Tracing this passage back to my previous analyses of *Glas*, I argue that the genetic structure of the trace-seed is precisely what allows us to dissociate natural life and the dissemination of the natural germ from their determination within the Hegelian system (namely, the book of life) and to think of them otherwise. Perhaps, as the element of another understanding of genesis (of another text). I put this hypothesis to the test in the next chapter where I examine an essay by Derrida that is contemporary with his lecture course on *La Famille de Hegel*.

5

HEGELIANISM III

The Genetic Programme

> Beginnings are always difficult in all sciences.
>
> —Marx, *Capital: A Critique of Political Economy*, 89[1]

In 1972, Derrida devotes his introductory essay to *Dissemination*, entitled "Outwork, Prefacing [*Hors livre, préfaces*]" (1972), to what he designates as the inescapable question of the textual preface. It is worth remarking from the outset that his notes on the preface do not apply to a classical or restricted concept of text, to a philosophical text among others, the Hegelian text, in this case. Rather, they refer to a concept of text that is modern and general, to the Hegelian text as the text of which there is no outside, the text of history and life, or, as Derrida calls it in *Glas*, the "book of life." For this reason, the question of the textual preface is also, for Derrida, the question of the genetic programme. It has to do with the difficult beginning of history and life, with the genesis of the logos as well as of the living. Through a strategic reading of the Hegelian concept of the textual preface and thus of the relationship between the preface and the general text, Derrida brings to light another thinking of the general text, the thinking of dissemination, in light of which we can read Hegel's prefaces and introductions, as well as his own preface. Within these coordinates, he also engages with the concept of the genetic programme as it is set up in the biological thought of his time. He demarcates dissemination from a formalist understanding of the biological genesis, which he traces back to the Hegelian concept of the preface. I argue that Derrida identifies this understanding with the concept of the genetic programme that the French molecular biologist François Jacob formalizes in *The Logic of Life* (1970). As we will see, a more or less explicit play of references to the introduction written by Jacob for the aforementioned volume, entitled "The Programme," undergirds Derrida's preface. On this point dissemination intersects post-*genetic* understandings of the biological genesis that are developed in the same years, among others, by the French biophysicist Henry Atlan.[2] Finally, I propose reading "Outwork, Prefacing" as a non-Hegelian and non-*genetic* protocol on beginning in general.

THIS IS A PROTOCOL

Derrida's text begins by remarking that it is not a preface, if we understand the latter from a traditional perspective—namely, as a pre-text carrying the instructions that will be unfolded in the text and thus constructing the originary relationship between the text itself and the subject that accompanies and protects it (its father). This traditional understanding of the preface recalls the concept of the genetic programme (or code), as a contemporary biologist such as Jacob elaborates it in *The Logic of Life*. Derrida's language seems to suggest this resonance, which I further develop in the analyses that follow. "Hence this is not a preface," he writes, "at least not if by preface we mean a table, a *code* [my emphasis], an annotated summary of prominent signifieds, or an index of key words or of proper names" (Derrida 1981, 8). By adopting the perspective of the preface itself (and thus of the genetic programme), Derrida describes the movement by which an author constituted a posteriori reappropriates the text as its own product. This movement consists in anticipating and thus mastering what is yet to come by reducing it to a present spectacle before the incorruptible understanding of a Leibnizian God. Derrida writes:

> From the viewpoint of the fore-word, which recreates an intention-to-say after the fact, the text exists as something

> written—a past—which, under the false appearance of a present, a hidden omnipotent author (in full mastery of his product) is presenting to the reader as his future. Here is what I wrote, then read, and what I am writing that you are going to read. After which you will again be able to take possession of this preface which in sum you have not yet begun to read, even though, once having read it, you will already have anticipated everything that follows and thus you might just as well dispense with reading the rest. The *pre* of the preface makes the future present, represents it, draws it closer, breathes it in, and in going ahead of it puts it ahead. The *pre* reduces the future to the form of manifest presence. (Derrida 1981, 7)

Derrida formulates the first thesis of dissemination by precisely dissociating the latter from the perspective of the preface. Dissemination is marked by a certain resistance of the text to reappropriation by *its* author and to the reduction of it to *its* meaning, which the preface would put to work. Above all, it requires another thinking of the preface, a thinking of the resistance of the preface to its traditional determination, and thus another thinking of the genetic programme. "One of the theses inscribed within dissemination," he affirms, "is precisely the impossibility of reducing a text as such to its effects of meaning, content, thesis, or theme." Rather than impossibility, he remarks, so long as prefaces are commonly written to this purpose, "perhaps" we should speak of "resistance" or "*restance*" (7–8). Therefore, dissemination may be understood, in the first place, as a thinking of the remains of what is commonly written as a preface. Within the framework of dissemination, Derrida suggests replacing the preface with a protocol: "A certain protocol *will have*—destroying this future perfect—*taken up* [my emphasis] the pre-occupying place of the preface" (8). In other words, the future perfect of the structure of the protocol will have resisted the reduction of the text to presence and to the originary relation to its author and meaning. Derrida makes this explicit in a footnote where he accounts for the protocol as the very structure of a supposedly originary logos.[3] From the perspective of dissemination, this structure is the trace-seed: the genesis of the logos as well as of the living—namely, of the logos-*zōon*—that I have been tracking in this book.

> In place of this discursive anticipation, the notion of "protocol" substitutes a textual monument: the first (proto-) page *glued* (*kollon*) over the opening—the first page—of a register or set of records. In all contexts in which it intervenes, the protocol comprises the meanings of priority, formula (form, pharmacopoeia), and writing: pre-scription. And through its "collage," the *protokollon* divides and undoes the inaugural pretention of the first page, as of any *incipit*. Everything, then, begins—this is a law of dissemination—doubled by a "facing." (8)

The resistance/restance of the preface, its protocollary structure, is even more noteworthy if we think with Derrida that "prefaces . . . have always been written, it seems, in view of their own self-effacement" (9).[4] The pages that follow can be read in light of the question of remains. While philosophy, as a certain modern and general understanding of the text, does not reckon with them, dissemination clings to them in view of thinking of the beginning and genesis of history and life otherwise. From the perspective of the preface, beginning and genesis are programmed to die, like the Hegelian concept of nature, or like the metaphor. As we see below, the irreducible structure of the protocol allows Derrida to think of a second death that the preface, like nature and metaphor, bears within itself.[5]

THE LOGIC TEXT

These initial remarks are put to the test through the examination of the Hegelian concept of the preface. Derrida starts observing that Hegel understands the preface from "the point of view . . . of the science of logic" (8)—that is, of the logic text of philosophical exposition. "Philosophical process," he remarks, "acts for itself as its own presentation, as the very domestic retinue [*domesticité*] of its own exposition (*Darstellung*)" (10). It is not by chance that a reference to "*domesticité*" recurs here. As we know, Derrida had adopted the term to account for the idea of spiritual life that grounds the organization of the Hegelian system and, at the same time, describes its self-development. The logic text thus must be understood as the book of life, which consists of the metaphorical exchanges between the ontological regions of the system, at the same time as it sublates these regions. It is from the perspective of the logic text, Derrida continues, that "Hegel . . . disqualifies the preface" (9).[6]

As pointed out in the preface to the *Phenomenology of Spirit*, philosophical exposition does not need the preface since it does not relate to an external and particular object. Rather, as Derrida explains, the thing itself is spontaneously produced and expressed in it.

> Philosophical exposition has as its essence the capacity and even the duty to do without a preface. This is what distinguishes it from empirical discourses (essays, conversations, polemics), from particular philosophical sciences, and from exact sciences, whether mathematical or empirical. . . . The preface to a philosophical work thus runs out of breath on the threshold of science. It is the site of a kind of chit-chat external to the very thing it appears to be talking about. This gossipy small talk of history reduces the thing itself (here the concept, the meaning of thought in the act of thinking itself and producing itself in the element of universality) to the form of a particular, finite object, the sort of object that determinate modes of knowledge—empirical descriptions or mathematical sciences—are incapable of producing spontaneously through their own workings and must therefore, for their part, introduce from the outside and define as a given. (Derrida 1981, 10)[7]

Therefore, the logic text relates to the preface as to any other regional discourse, by determining the latter's position in the system and thus sublating it. "Isn't the preface both negated and internalized in the presentation of philosophy by itself, in the self-production and self-determination of the concept?" (14), Derrida wonders. However, this does not prevent him from drawing attention to what remains of the preface, from holding on to this resistance and, therefore, sketching out another thinking of the preface itself.

I interrupt for a moment the analysis of the Hegelian concept of the preface to come back to the perspective of the science of logic that Derrida evokes in this passage. A few pages prior, in a footnote related to his initial remarks, Derrida seems to summarize this perspective by focusing on the following passage from the *Science of Logic*: "Difference as such is already implicitly contradiction" (Hegel 2010, 374).[8] According to Derrida, "the movement by which Hegel determines difference as contradiction," that he identifies with the quoted passage, "is designed precisely to make possible the ultimate (onto-theo-teleo-logical) sublation [*la relève*] of difference" (Derrida 1981, 6–7). We know that, for Derrida, the onto-theo-teleo-logical is the element of the Hegelian text, the infinite germ of spiritual life, which, at once, regulates the metaphorical exchanges among the regions of the text and embodies the latter's self-unfolding. Therefore, the ultimate reduction of difference to contradiction accounts for the *parti pris* of the book of life: the determination of natural and biological life on the basis of the self-reproductive and incorruptible life of the spirit, which Derrida has called the *logos spermatikos* since "Force and Signification." He contrasts this perspective with that of differance, which consists in the dissemination of the trace-seed as the minimal structure of the logos as well as of the living and thus in the syntactical and tropic movements that, as I demonstrate in the previous chapters, the Hegelian text has always already presupposed. To follow Derrida's interpretation of the movement described in the Hegelian sentence, we should take account of the two concepts of general text that are at stake here: the logic text (namely, the book of life) and the text of differance (dissemination). Derrida describes the latter as follows:

> Differance—which is thus by no means dialectical contradiction in this Hegelian sense—marks the critical limit of the idealizing powers of sublation [*la relève*, my translation] wherever they are able, directly or indirectly, to operate. Differance *inscribes* contradiction, or rather, since it remains irreducibly differentiating and disseminating, contradiction*s*. (Derrida 1981, 6)

THE TWO DEATHS OF THE PREFACE

In the Hegelian explanation of the necessity of the preface, as Derrida interprets it in the subsequent pages of "Outwork, Prefacing," we find the traces of the nonformalist conception of writing that Derrida had developed in "Force and Signification" in the wake of Husserl's understanding of the absolutely constitutive character of geometrical productions. "Meaning [*sens*] must await being said or written in order to inhabit itself, and in order to become, by differing from itself, what it is: meaning" (Derrida 1978, 10), Derrida writes in his earlier essay. Then, he concludes: "This is what Husserl teaches us to think in *The Origin of Geometry*" (10). As we will see, in Derrida's Hegel, the preface takes on the place of writing, and sense (understood as absolute knowledge, concept, etc.) must await it in order to become what it is. Derrida points out that "the prefatory moment is necessarily opened up by the critical gap between the logical or scientific development of philosophy and its empiricist or formalist lag" (Derrida 1981, 11). This irreducible delay is attached to what, in the preface to the *Phenomenology*, Hegel identifies as the "external necessity" of the self-presentation of the concept: time as the existence (*Dasein*) of the concept itself. The delay is explained by "the gap between this formal notion of time, the general matrix in which the concept is present, and the empirical or historical determination of time, that of our time, for example" (12). Derrida is reading the following text from Hegel's preface:

> But the external necessity, so far as it is grasped in a general way, setting aside accidental matters of person and motivation, is the same as the inner, or in other words it lies in the shape [*Gestalt*] in which time sets forth the sequential existence of its moments [*wie die Zeit das Dasein ihrer Momente vorstellt*]. To show that now is the propitious time [*an der Zeit*] for philosophy to be elevated to the status of a Science would therefore be the only true justification of any effort that has this aim, for to do so would demonstrate the necessity of the aim, would indeed at the same time be the accomplishing of it. (Hegel 1977, 3–4)

Derrida draws attention to the fact that time is out of joint, unequal to itself, etc., and the preface comes to repair it. The latter is required precisely to fasten together the concept and the time of its existence. The space and time taken by the preface work as a necessary supplement to the time of the existence of the concept. As Derrida puts it in a Shakespearean lexicon, which comes back in his later work:

> But since our time is not exactly, not simply propitious for such an elevation (*Erhebung*), since it is not yet quite the right

time (*an der Zeit*), since the time, at any rate, is not equal to itself, it is still necessary to prepare it and make it join up with itself by didactic means. . . . A certain spacing between concept and being-there, between concept and existence, between thought and time, would thus constitute the rather unqualifiable lodging of the preface. (Derrida 1981, 12)

When the work of the preface is accomplished and the lag is repaired, the preface effaces itself according to its internal programme, just as much as nature commits suicide in order to liberate spirit.[9] This is what Derrida suggests, marking the point where his nonformalist thinking of the preface and, more generally, of genesis, diverges from Hegel's. Here we touch upon the point of the resistance of the preface and genesis, as well as upon the point of departure for dissemination. It is not by chance that, in view of situating the preface, Derrida understands it as an inscription on the place of *khōra*—namely, as a tropic or syntactical movement that precedes the determination of philosophical oppositions. He remarks that the preface stands for a sort of third term beside form and content, signifier and signified, and so forth, which cannot be thought on the basis of these oppositions and thus within the limits of philosophy. It carries out a function of announcing that turns the self-development of the concept into a movement of anticipation and delay. As Derrida observes,

> What is the status of this third term which . . . is neither a pure form, completely empty, since it announces the path and the semantic production of the concept, nor a content, a moment of meaning, since it remains external to the logos of which it indefinitely feeds the critique, if only through the gap between ratiocination and rationality, between empirical history and conceptual history? If one sets out from the oppositions form/content, signifier/signified, sensible/intelligible, one cannot comprehend the writing of a preface. But in thus *remaining*, does a preface exist? Its spacing (the preface to a rereading) diverges in (the) place of the *khōra*. (1981, 15–16)

Derrida highlights the point of departure of dissemination from the Hegelian thinking of the preface by examining the status of the Introduction to the *Science of Logic*. He focuses on a passage from the first page in which Hegel seems to reformulate the understanding of the preface that he had elaborated earlier in the Preface to the *Phenomenology of Spirit*. Logic admits neither an anticipation nor a genesis, so long as the genesis of the concept coincides with the completion of logic itself. Therefore, the introduction can be justified only as an anticipation that responds to the empirical or formal delay of our times. Hegel writes:

> Logic, therefore, cannot say what it is in advance, rather does this knowledge of itself only emerge as the final result and completion of its whole treatment. Likewise, its subject matter, thinking or more specifically conceptual thinking, is essentially elaborated within it; its concept is generated in the course of this elaboration and cannot therefore be given in advance. What is anticipated in this Introduction, therefore, is not intended to ground as it were the concept of logic, or to justify in advance its content and method scientifically, but rather to make more intuitive, by means of some explanations and reflections of an argumentative and historical nature, the standpoint from which this science ought to be considered. (Hegel 2010, 23)

In his explication of the text, Derrida emphasizes the relationship between logic and introduction. Logic—namely, the logic text, domesticity itself, etc.—is the element that regulates at the same time as sublates the organization of the sciences. As Derrida explains, "Even in its exordium [beginning, genesis, and so forth] logic moves already in the element of its own content and need not borrow any formal rules from any other science" (Derrida 1981, 19). Therefore, understood as analogous to any regional science, the introduction has already taken place in logic and thus has already been programmed to die. "Hegel denies the logical character of his Introduction in conceding that it is but a concession, that it *remains*, like classical philosophy, external to its content, a mere formality designed to remove itself on its own initiative" (19), Derrida points out. The introduction is "constructed" according to "the Hegelian values" (20) of speculative dialectics.[10] From the perspective of the logic text, that is, of the self-engendering and incorruptible germ of spiritual life, every anticipation or genesis is determined on the basis of the text itself and thus as a preface or an introduction. Derrida supposes that the logic text occurs in the present and has no duration. The spacing of the preface and the construction of the author and meaning are tied together and taken up as moments in the continuity and homogeneity of this text. Therefore, the logic text consists in the relationship of the logos (philosophical exposition) and its subject (absolute knowledge). Derrida argues:

> The signifying *pre-cipitation*, which pushes the preface to the front, makes it seem like an empty form still deprived of what it wants to say; but since it is ahead of itself, it finds itself predetermined, in its text, by a semantic *after-effect*. But such indeed is the essence of speculative production: the signifying precipitation and the semantic after-effect are here *homogeneous* and *continuous*. Absolute knowledge is present at the zero point of the philosophical exposition ... This point of ontoteleological fusion reduces both precipitation and after-effect to mere appearances or to sublatable negativities. (20)

The zero point of onto(-theo-)teleological fusion is the book of life—namely, the organization of the Hegelian system and its self-unfolding. Here we measure the proximity and distance of Hegel's concept of text from the *modern* concept of the general text (which reminds us of the *modern* concept of writing that Derrida elaborates in "Force and Signification"). On the one hand, the logic text not only regulates the relations among regional texts (preface, genesis, etc.) but has already been at work in each of them, as it also describes the self-development of the system. Derrida accounts for this general concept of the text by echoing a passage from *Of Grammatology* where he conceives of the scriptural structure of the pro-*gramme* as the element of history and life, from the genetic programme of the cell to the programme of cybernetic machines.[11] I return later to this point as Derrida conjures up again this passage from *Grammatology*. Apropos of Hegel's general text, Derrida remarks that "nothing precedes textual generality absolutely" and thus "there is no preface, no program, or at least any *program* is already a pro*gram*, a moment of the text, reclaimed by the text from its own exteriority" (20). Therefore, the programme seems to be the minimal structure of the logic text. However, on the other hand, this text is seen as present, without duration, and related to a constructed author and meaning—which Derrida calls "sense" by linking the logic text to the Hegelian solution of the equivocity between natural and philosophical language that is at stake in the natural-speculative word *Sinn*. He thus demarcates the logic text from the modern concept of the general text, which constitutes the minimal structure of the logos as well as of the living, the syntactical and tropic text of traces-seeds. "But Hegel brings this generalization about by saturating the text with meaning [*sens*]," Derrida explains, "by teleologically equating it with its conceptual tenor, by reducing all absolute dehiscence between writing and wanting-to-say (*vouloir-dire*), by erasing a certain occurrence of the break between anticipation and recapitulation" (20).[12] As we know from "Force and Signification," and in the wake of a certain reading of Husserl's *Origin*, but, also, as Hegel explains in his preface to the *Phenomenology*, this break is the irreducible condition for sense in general (namely, for history and life) to become what it is and thus to differ from itself. This break is precisely what makes every *pro*gramme into a pro*gramme*, every beginning or genesis into a trace or a natural seed. At this point, Derrida spells out the modern concept of the general text that arises from the perspective of dissemination. "If the preface appears inadmissible *today* [my emphasis] . . ." (20): he begins by highlighting the change of perspectives. Prefaces (and, more generally, beginnings) are inadmissible so long as we conceive of them from the perspective of the logic text; they are constructions of speculative dialectics and are solidary with the latter's values. At the same time, we can retain them insofar as they account for the very structure of the signifying precipitation and thus they have already been generalized as a programme (or a trace-seed). In the following passage, Derrida formulates a thinking of the preface that is neither thematic nor formalist:

> If the preface appears inadmissible today, it is on the contrary because no possible heading can any longer enable anticipation and recapitulation to meet and to merge with one another. To lose one's head, no longer to know where one's head is, such is perhaps the effect of dissemination. If it would be ludicrous today to attempt a preface that really was a preface, it is because we know semantic saturation to be impossible; the signifying precipitation introduces an excess facing (*un débord*) . . . that cannot be mastered; the

> semantic after-effect cannot be turned back into a teleological anticipation and into the soothing order of the future perfect; the gap between the empty "form" and the fullness of "meaning" is structurally irremediable, and any formalism, as well as any thematicism, will be impotent to dominate that structure. They will miss it in their very attempt to master it. The generalization of the grammatical or the textual hinges on the disappearance, or rather the reinscription, of the semantic horizon, even when—especially when—it comprehends difference or plurality. In diverging from polysemy, comprising both more and less than the latter, dissemination interrupts the circulation that transforms into an origin what is actually an after-effect of meaning. (Derrida 1981, 20–21)

Generalization supposes a second death of the preface, as well as of genesis, which does not unfold through their sublation in the logic text but, rather, through the inscription within the general text of history and life and thus through dissemination. The final sentence of this passage recalls the distinction between origin and genesis that, for Derrida, has been at stake in philosophy's departure from mythology since Plato.[13] If the logic text described so far thinks of genesis either as programmed to die or as origin, dissemination is a thinking of the genetic structure in general, of the general structure of linguistic as well as biological genesis.

Referring to a text already mentioned in his *Introduction* to Husserl's *Origin* and in "Force and Signification," precisely in relation to the same issue, here Derrida evokes, once more, Feuerbach's reading of the Hegelian understanding of beginning in *Towards a Critique of Hegel's Philosophy*. "It should be specified," Derrida remarks, "that Feuerbach had already examined in terms of writing the question of the Hegelian *presupposition* and of the textual residue" (29). In other words, Feuerbach interpreted beginning as writing, as an inscription that announces what is yet to come (namely, sense), and cannot be fully reappropriated by a constructed meaning. A tension remains unresolved between philosophical exposition and its beginning. In the passage quoted by Derrida, Feuerbach seems to argue that the text of the *Logic* and, more generally, the logic text, necessarily bear within themselves the remains of their beginning, a protocollary rest, which only another thinking of beginning (such as dissemination) could take up.[14]

Derrida articulates the two deaths of the preface, as well as of genesis, by having recourse to the figure of "an uneven chiasmus" (35). On the one hand, what Hegel rejects as a preface has already been shown to be more than this: it is the general structure of genesis. On the other hand, dissemination dismisses the preface as it is commonly understood, or, in other words, as what constructs the author and the meaning that reappropriate a text a posteriori. On the one hand, the preface names a structure that cannot be thought according to the values of speculative dialectics and thus resists the logic text. It lends itself to an altogether different thinking of the general text. This is the first arm of the chiasm:

> In Hegel's reason for disqualifying the preface (its formal exteriority, its signifying precipitation, its textuality freed from the authority of meaning or of the concept, etc.), how can we avoid recognizing the very question of writing, in the sense that is being analysed here? The preface then becomes necessary and structurally interminable, it can no longer be described in terms of a speculative dialectic: it is no longer merely an empty form, a vacant significance, the pure empiricity of the non-concept, but a completely other structure, a more powerful one, capable of accounting for effects of meaning, experience, concept, and reality, reinscribing them without this operation's being the inclusion of any ideal "*begreifen.*" (Derrida 1981, 35)

On the other hand, dissemination does not dismiss the preface for the same reasons as Hegel's. Rather, it accounts for the structure of the preface and, more generally, of the beginning that resists the logic text. In other words, dissemination reckons with the general text in which this structure is situated. Therefore, it dismisses a formalist as well as thematic understanding of the preface: "What in our eyes today makes those prefaces [the Hegelian prefaces] appear . . . contrary to *the necessity of the text* [my emphasis], written in an outworn rhetoric suspect in its reduction of the chain of writing to its thematic effects or to the formality of its articulations" (35).

THE PREFACE IS THE NATURE OF THE LOGOS

I pointed out that the question of the general text is intimately linked to that of the organization of knowledge and of the relations between philosophy and the sciences (for instance, the life sciences) in the text itself. For Derrida, the Hegelian concept of the general text—namely, the logic text, hinges on the concept of life as the life of the concept—the self-engendering and incorruptible germ of spiritual life—and takes the latter as the figure of its organization and self-development. Dissemination puts this Hegelian scheme into question by calling for another thinking of the general text, in which linguistic as well as biological geneses are inscribed. As Derrida suggests in a footnote on Marx's criticism of Hegel's concept of the preface, "It is the whole scheme of the subordination of the sciences, and then of the regional ontologies, to a general or fundamental onto-logic that is perhaps here being thrown into confusion" (33). Derrida brings this confusion to the fore when he interprets the relationship between the Hegelian *Encyclopedia Science of Logic* (1817, part I) and its preface, in light of the relationship between concept and nature as it is developed in the *Encyclopedia Philosophy of Nature* (part 2). He understands the *Encyclopedia* on the basis of the Hegelian concept of the general text and thus of the logic text of philosophical exposition. The *Encyclopedia* unfolds the very solution of the equivocity between natural and spiritual life—that is, the Hegelian syllogism of life and the speculative path of the metaphor. This analysis of the relationship between the *Encyclopedia* and the preface focuses on sections 1 and 10 in the "Introduction" to the *Encyclopedia Science of Logic*, which Derrida interprets from the perspective of the science of logic elaborated above. As he highlights, section 1 recalls that the *Encyclopedia* demarcates itself from the other sciences as it lacks "the ability to presuppose both its *object* . . . and *method* of knowing" (Hegel 2010b, 28). Hence, "the difficulty of making a beginning" . . . "since a beginning is *something* immediate," Hegel continues, "and as such makes a presupposition, or rather it is itself just that" (28). Derrida observes that the encyclopedia "must produce, out of its own interiority, both its object and its method" (Derrida 1981, 47), as the general text according to Hegel—namely, the text of the *Logic*, does. Section 10 marks the distinction between "the thinking operative in the philosophical manner of knowing" and "a preliminary explication," which would necessarily be non-philosophical and "could not be more than a web of presuppositions, assurances, and formal reasoning, a web, that is, of casual assertions against which the opposite could be maintained with equal right" (Hegel 2010b, 38). Derrida reformulates this relationship between philosophy (and thus the text of the *Encyclopedia*) and a preliminary exposition by referring to the relationship between natural and spiritual life and thus in terms of the syllogism of life. The *Encyclopedia* embodies the self-reproductive and eternal germ of spiritual life out of which we should think of the natural germ as its natural image (or a metaphor) and thus as programmed to die. Derrida writes:

> Engenderer and consumer of itself, the concept relieves (*relève*) its preface and plunges into itself. The Encyclopedia gives itself birth. The conception of the concept is an auto-insemination. This return of the theological seed to itself internalizes its own negativity and its own difference to itself. The Life of the Concept is a necessity that, in *including* the dispersion of the seed, in making that dispersion work to the profit of the Idea, *excludes* by the same token all loss and all haphazard productivity. The exclusion is an inclusion. In contrast to the seminal difference thus repressed, the truth that speaks (to) itself within the logocentric circle is the discourse of what *goes back to the father*. (Derrida 1981, 48)

However, the natural germ is more than what the *Encyclopedia* calls by this name. The natural germ resists and offers itself to another thinking of the general text: it accounts for the genetic structure in general. Derrida speaks about seminal differance by looking at the natural germ from the perspective of dissemination. Furthermore, in a related footnote, he suggests that the problem of the equivocity between natural and spiritual life is also at stake in the *Encyclopedia* and, once again, Hegel eludes it by holding on to the *parti pris* of life as the spiritual germ. This *parti pris* constitutes the onto-theological figure of the *Encyclopedia*: it accounts for the metaphorical exchange between regional discourses in the text as well as for the latter's self-development and presence. The footnote reads:

> Life, the essential philosophical determination both of the concept and of the spirit, is necessarily described according

to the general traits of vegetal or biological life, which is the particular object of the philosophy of nature. This analogy or this metaphoricity, which poses formidable problems, is only possible following the *organicity* [my emphasis] of encyclopedic logic. (Derrida 1981, 48)

It is from this perspective that we may read the whole *Encyclopedia* as the syllogism of life, "as the life of the spirit as truth and death (termination) of the natural life that bears within itself, in its finitude, 'the original disease . . . and the inborn germ of death'" (48). The preface, as a certain kind of beginning, is classified according to the *parti pris* of the spiritual germ, and thus it is programmed to die and efface itself. For this reason, Derrida raises the following questions: "Is the preface the nature of logos? The natural life of the concept?" (48). In conclusion, Hegel does not take into account what of the preface and, more generally, of the natural germ—that is, seminal differance—resists the general text of the *Encyclopedia*. He does not take into account what becomes reconsidered, from the perspective of dissemination, as the minimal condition for genesis.[15]

THE TAIN OF THE MIRROR

The relationship between natural and spiritual life is implicit in the formulation of dissemination that Derrida offers a few pages above, as he recalls the concluding paragraphs of the *Philosophy of Nature* in the *Encyclopedia*. Once more, Derrida remarks that the preface (and beginning), as the inscription of a sense that is not constituted yet, escapes the hold of the Hegelian and philosophical understanding of the preface (and beginning) as a merely empirical or formal fact. Dissemination enters the scene precisely as it engages with these remains and conceives of them as the genesis of the logos and the living. Derrida writes:

> The breakthrough toward radical otherness (with respect to the philosophical concept—of the concept) always takes, *within philosophy*, the form of a posteriority or an empiricism. But this is an effect of the specular nature of philosophical reflection, philosophy being incapable of inscribing (comprehending) what is outside it otherwise than through the

appropriating assimilation of a negative image of it, and dissemination is written on the back—the tain—of that mirror. (1981, 33)

If empiricism is the "mask" that philosophy applies to the "heterological breaching (*frayage*)" in view of mastering it, dissemination dissociates itself from empiricism as it uncovers this breaching under its philosophical mask and turns it into the genetic structure of history and life. As anticipated, the passage quoted above seems to recall the last page of the *Philosophy of Nature* (Organics, section 376). Here Hegel observes that "the difficulty of the Philosophy of Nature" lies in the fact that "the material element is so refractory toward the unity of the Notion" and thus "spirit has to deal with an ever-increasing wealth of detail" (Hegel 1970, 444). However, he invites Reason to have "confidence in itself," "confidence that . . . the veritable form of the Notion which lies concealed beneath Nature's scattered and infinitely many shapes, will reveal itself to Reason" (444–45). Therefore, he finds the purpose of the philosophy of nature to be the subjugation of the polymorphy of natural life. He accounts for this purpose by having recourse to a series of terms that come back in Derrida's formulation of dissemination in relation to empiricism. The purpose of the philosophy of nature, Hegel concludes, "has been to give a picture of Nature in order to subdue this Proteus: to find in this externality only the mirror of ourselves, to see in Nature a free reflex of spirit" (445). Now, dissemination is concerned with those traces-seeds that resist the philosophical mask, the general text, and the syllogism of life, and constitute the general structure of genesis. Dissemination is thus inscribed on the back of the Hegelian mirror of nature. Derrida comments on this passage once again in *Glas*, as he brings to an end his close reading of the transition from the philosophy of nature to the philosophy of spirit in the *Encyclopedia*. Here he draws attention to the figures of the reflex and the mirror, and he remarks that the philosophical character of nature constructed by Hegel is complicitous with the unfolding of the encyclopedia and thus wants to be overcome by spirit. As the text from *Glas* reads, "The Proteus had strictly to be subjugated (*diesen Proteus zu bezwingen*). Nature will have asked for nothing else: 'The purpose of nature is to kill itself'" (Derrida 1986, 117).

The question of the encyclopedia as a general text is taken further through the reference to the encyclopedic project elaborated by

Novalis (edited in France by Maurice de Gandillac in 1966). According to Derrida, this project "explicitly poses the question of the form of the total book as a written book: an exhaustive taxonomical writing, a hologram that would order and classify knowledge" (Derrida 1981, 50). It does this by "*giving place* to literary writing" (50), and thus by supposing that the general text encompasses what writing stands for: an inscription of sense, a signifying precipitation, a disseminated germ, etc. This reading of Novalis's project focuses on the question of the relationship between the preface and the general text of the encyclopedia. Derrida reads Novalis in the wake of Hegel: the encyclopedia would consist of the organization and development of knowledge according to the onto-theological principle of the spiritual germ and thus of the *logos spermatikos*. Derrida writes:

> The question of the genetic pro-gram or the textual preface can no longer be eluded. Which does not mean that Novalis does not *in the final analysis* reinstall the seed in the *logos spermatikos* of philosophy. Postface and preface alike will return to the status of Biblical moments. Comprehended *a priori* within the volumen. (50)

The "Biblical moments" in this passage signal the reduction of writing and seminal differance to determinations of the *logos spermatikos*. Derrida interprets the encyclopedia of Novalis as a simultaneous spectacle before a divine understanding and thus as the self-development of the incorruptible germ of spiritual life. As he suggests at the end of a long quotation from Novalis, "History itself is prescribed. Its development, its violence, even its discontinuities should not disconcert this musical volume" (51); and, a paragraph later, "It is a Bible, then, as tabular space but also as seminal reason explaining itself" (52). However, according to Derrida, dissemination too, as a general text, *explains itself*, but neither as a prescribed history nor as a self-reproductive and incorruptible germ.[16] In a related footnote, he develops the reference to the philosophical tradition of the *logos spermatikos* by dissociating the latter from dissemination. He identifies this tradition as "the philosophy of the seed" described by Gaston Bachelard in *The Formation of the Scientific Mind* (1938), whose key features are "self-return" and "substantialism" (50).[17] Dissemination is the name given to a project that he has started elaborating in "Force and Signification" (as I suggest in this book) and further developed in *Of Grammatology*, where he thinks of writing as the general structure of genesis and thus as the element of history and life. "It is rather a question of broaching," he observes, "an articulation with the movement of *genetic science* and with the *genetic movement of science* [my emphasis], wherever science should take into account, more than metaphorically, the problems of writing and difference, of seminal difference" (50). Dissemination is the thinking of the general text of history and life whose elements are writing and seminal difference. Once again, Derrida borrows from the biological thought of his time the scriptural determination of biological genesis—namely, the concept of the genetic programme—and he reinterprets it from the perspective of the general text of dissemination. For this reason, I suggest, he concludes the examined footnote by recalling Freud's wait for the biological discovery of an organic structure that would confirm his psychological hypotheses. What is the biological structure of the element of dissemination, of writing, seminal difference, programme, etc.? Returning to the hypothesis of the self-explanation of the general text of dissemination, Derrida (1981) argues that this structure resists formalism (as well as thematicism) and this resistance can be formalized. He writes:

> As the heterogeneity and absolute exteriority of the seed, seminal difference does constitute itself into a program, but it is a program that cannot be formalized. For reasons that *can* be formalized. The infinity of its code, its rift, then, does not take a form saturated with self-presence in the encyclopedic circle. It is attached, so to speak, to the incessant falling of *a supplement to the code* [*un supplement de code*]. Formalism no longer fails before an empirical richness but before a queue or tail. Whose *self*-bite is neither specular nor symbolic. (51)

This is perhaps an inescapable passage for any explorations of "Outwork, Prefacing" and, more generally, of Derrida's early work. He describes the structure of genesis (seed, precipitation, heterological breaching, etc.) as a programme—what he had called "inscription" in "Force and Signification." However, this programme is not simply reduced, as a purely empirical or formal element, to the general text of an incorruptible germ or seminal reason, such as the logic text or the encyclopedia. As the reference to the supplement of code suggests, the programme is situated within the dimension of the tropic and syntactical chains that constitute the general text of its textuality, and thus it refers to other chains, indefinitely. The supplement of code accounts for the inability of any text to set itself free from its textuality and to develop itself as a spiritual germ or a seminal reason.[18] I suggest reading these lines as Derrida's reformulation of the biological concept of the genetic programme—the properly called genetic programme—in light of his understanding of genesis.[19]

In *The Logic of Life*, Jacob seems to elaborate the conceptual framework that remains implicit in the examined passage from "Outwork, Prefacing." I refer to chapter 5, dedicated to "The Molecule," where Jacob focuses on the way molecular biology revolutionized the understanding of biological heredity. He explains that the sequence of the nucleic acid (DNA), in which the hereditary characters of a species—namely, the genes—are inscribed, constitutes the programme for the synthesis of proteins in the cell and thus for the development of the living organism. Within this framework, Jacob determines the relationship between the nucleic-acid sequence and the synthetized molecules of proteins as a relationship of transformation between two texts or chains of symbols:

> The nucleic-acid sequence determines the order of the protein sub-units. This is a unidirectional process: the transfer of information always goes from nucleic acid to protein, never in the other direction. But whereas the combinative system of nucleic acid uses only four chemical symbols, that of protein uses twenty. The activity of the gene, the execution of instructions for protein synthesis therefore requires the univocal transformation of one system of symbols to the other. (Jacob 1973, 275)

As Jacob points out a few pages later, we can also understand this relation of transformation in terms of translation. It takes place within the framework of the general text elaborated by Norbert Wiener, the founder of cybernetics, and put to the test, at the time, in the ways radio, television and secret service worked. "By isomorphous transformation according to a code," Jacob observes, "such a structure [a message] can be *translated* [my emphasis] into another series of symbols. It can be communicated by a transmitter to any point on the globe where a receiver reconstitutes the message by reverse transformation" (252).[20] From this perspective, he concludes by quoting Wiener, "there is no obstacle to using a metaphor 'in which the organism is seen as a message'" (252).[21]

A NON-*GENETIC* THINKING OF GENESIS

In this final section, I further elaborate the relationship between the thinking of genesis formulated in Derrida's preface and the biological theory that he alludes to as a "genetic science" in the aforementioned footnote from "Outwork, Prefacing." As explained, Derrida addresses the question of the genetic programme from the perspective of dissemination and thus understands the programme as a structure that is neither thematic nor formal, but the inscription of a sense yet to come or of the absolutely other. From this perspective, the preface to *Dissemination* can be read as a critical engagement with Jacob's introduction to *The Logic of Life*, entitled "The Programme," so long as the latter holds on to a formalist understanding of the programme and thus falls back into the tradition of the philosophy of the seed and of the *logos spermatikos*. While Jacob's introduction elaborates the concept of a programme that explains itself as a seminal reason, Derrida's preface marks the limits of a philosophical understanding of the preface. The latter is inscribed on the back of the general text of philosophy. Therefore, the unavoidable question of the preface is also the question of the genetic programme. My hypothesis is that, with his preface, Derrida takes sides in contemporary debate surrounding the conditions for biological genesis. "Outwork, Prefacing" is thus the protocol of a non-Hegelian and non-*genetic* understanding of genesis.[22]

A close reading of "The Programme" would not only show that Jacob could not understand what the programme stands for from the perspective of dissemination. It would also prove that Jacob's genetic programme constitutes the latest version of the mask that the philosophy of the seed has always placed upon writing and seminal difference. Therefore, Derrida's preface is written on the back of Jacob's "Programme" as well as, for instance, of Hegel's introduction to the *Science of Logic*. Unfolding his understanding of the programme, Jacob himself establishes a striking continuity between genetics and the philosophy of the seed described by Bachelard in *The Formation of the Scientific Mind*.

He argues that the concept of the programme allows biologists of the time to solve the paradox that the natural sciences have struggled with throughout their history—namely, the conflict between the physico-chemical explanation of the phenomena of life and the teleological design that seems to underlie them. Before referring to Claude Bernard's formulation of this paradox in the nineteenth century, Jacob articulates it as follows: "The opposition between the mechanistic interpretation of the organism on one hand, and the evident finality of certain phenomena, such as the development of the egg into an adult, or animal behaviour, on the other" (4). Therefore, Jacob's understanding of the programme remains within the limits of the logic and encyclopedic text—that is, of the self-explanation of a seminal reason. I quote the passage from Bernard's text commented by Jacob:

> Even if we assume that vital phenomena are linked to physico-chemical manifestations, which is true, this does not solve the question as a whole, since it is not a casual encounter between physico-chemical phenomena which creates each being according to a predetermined plan and design. . . . Vital phenomena certainly have strictly defined physico-chemical conditions, but at the same time they are subordinated and succeed each other in sequence according to a law laid down in advance; they are repeated over and over again in ordered, regular and constant manner, harmonizing with each other, with a view to achieving the organization and growth of the individual, animal or plant. There is a kind of pre-established design for each being and each organ, so that, considered in isolation, each phenomenon of the harmonious arrangement depends on the general forces of nature, but taken in relationship with the others, it reveals a special bond: some invisible guide seems to direct it along the path it follows, leading it to the place which it occupies. (Jacob 1973, 4)

After countersigning this passage ("not a word of these lines needs to be changed today: they contain nothing which modern biology cannot endorse," 4), Jacob argues that the concept of the genetic programme developed by molecular biology reconciles teleological evidences with scientific explanations (and thus the general text with the preface).

"However, when heredity is described as a coded programme in a sequence of chemical radicals," he writes, "the paradox disappears" (4). Jacob explains that this biological description of heredity as the transmission and execution of the instructions required to the development of the future organism is borrowed from cybernetics. These instructions, archived in the genetic programme, make the living be present as a future work and trace it back to a constructed teleological design. As Jacob observes, "Heredity is described today in terms of information, messages and code" so long as "what are transmitted from generation to generation [sequences of four chemical radicals contained in the genetic heritage] are the 'instructions' specifying the molecular structures: the architectural plans of the future organism" (1). A formulation of the programme follows, which can be read as the philosophical concept of programme whose limits are demarcated in Derrida's preface. Jacob explicitly determines his concept of programme within the limits of the *logos spermatikos*. The fact that he conceives of his general text as freed from the authority of an author and meaning does not imply a break with that tradition. He writes:

> In the chromosomes received from its parents, each egg therefore contains its entire future: the stages of its development, the shape and the properties of the living being which will emerge. The organism thus becomes the realization of a programme prescribed by its heredity. The intention of a psyche has been replaced by the translation of a message. The living being does indeed represent the execution of a plan, but not one conceived in any mind. It strives towards a goal, but not one chosen by any will. The aim is to prepare an identical programme for the following generation. (2)

Therefore, Jacob concludes that "with the application to heredity of the concept of programme, certain biological contradictions formerly summed up in a series of antitheses at last disappear," for instance, the antithesis between "finality and mechanism," or the one between "necessity and contingency" (5).

This philosophical understanding of genesis is confirmed in the chapter of the book dedicated to the doctrine of preformationism (and pre-existence), which Jacob presents as the moment within the history of the natural sciences when the biological paradox is resolved by resorting to a divine author or intention. According to Jacob, this is what merely differentiate preformationism from molecular biology within the tradition of the philosophy of the seed. In other words, the germ of preformationists as well as the genetic programme of molecular biologists are philosophical masks of genesis. In the section on "Preformation," in chapter 1 of *The Logic of Life* ("The Visible Structure"), Jacob explains that, in the seventeenth and eighteenth centuries, "the secret of the seed is hidden in the seed" to the extent that it is the only aspect of generation "accessible to their means of observation" (55). In this context, the theory of preformation is elaborated as follows:

> To maintain the continuity of shape, the "germ" of the little being to come has to be contained in the seed; it has to be "preformed." The germ already represents the visible structure of the future child, similar to that of the parents. It is the plan of the future living body, not potentially, in some active part of the seed from which the body of the little being is organized progressively, in the same way as a plan is carried out; but already materialized, like a miniature of the organism to come. It is like a scale model with all the parts, pieces and details already in position. The complete, although inert, body of the future being lies already waiting in the germ. Fertilization only activates it and starts it growing. (Jacob 1973, 57)

Jacob interprets the development of preformationism into the doctrine of pre-existence as a response to the question of teleology. From the perspective of dissemination, this development only uncovers the implicit presupposition of preformationism as well as, more generally, of the philosophy of the seed (including Jacob's concept of the genetic programme): the presupposition of a simultaneous spectacle before a Leibnizian God, which can be constructed only a posteriori, in the text of a preface. As Jacob remarks, "Only one solution was left: to consider that the germs of all organisms past, present or future, had always existed, that they had been formed at the time of Creation and were only awaiting the moment of activation by fertilization" (60). "This," he continues, "was the theory of the 'pre-existence' of germs" (60).

POSTSCRIPT

> If philosophy in fact has an "irreplaceable function," is it because nothing could replace it were it to die? I believe instead that it is always replaced: such would be the form of its irreplaceability. That is why the fight is never simply for or against Philosophy, the life or death, the presence or absence, in teaching, of Philosophy, but between forces and their philosophical instances, inside and outside of the academic institution.
>
> —Derrida, *Who Is Afraid of Philosophy:*
> *Right to Philosophy I*, 172[1]

In what follows, I push my analysis of dissemination a little further by highlighting a peculiar moment of Derrida's work, in 1974–1975, which constitutes a step forward in his elaboration of the minimal conditions for genesis. Derrida discovers that the germ of death—that is, the genetic structure of the living and of discourse—is irreducibly attached to a drive for mastery and domination and thus to a certain concept of power. From this discovery, it becomes explicit that politico-institutional bodies have always already been inscribed in the space of dissemination. To take account of this moment of Derrida's work, I discuss a group of texts presented in the lecture course *GREPH (Le concept de l'idéologie chez les idéologues français [The Concept of Ideology in the French Ideologues])*, delivered in 1974–1975, and partially published in *Du droit à la philosophie* (*Right to Philosophy*) in 1990. These texts were written on occasion of the school system reform proposed by the minister of education of the time, René Haby. They testify the struggle of the *GREPH* (*Group de recherche pour l'enseignement de la philosophie*), of which Derrida was one of the most active contributors, against the Haby reform. My analysis focuses on Derrida's interpretation of the teaching body as an institutional and political body driven by interest and power, tracing this interpretation back to Derrida's reading of Marx's understanding of sexual difference as the most general condition of ideology.

I.

In the essay "Where a Teaching Body Begins and how It Ends," which was originally published in 1975, Derrida starts by wresting the concept of teaching body—of the institutional body of philosophy teachers—away from any operation of naturalization and neutralization. He highlights "the forces and interests that, without the slightest neutrality, dominate and master . . . the process of teaching from within a heterogeneous and divided agonistic field wracked with constant struggle" (Derrida 2002b, 69). In other words, the teaching body is constituted by a set of forces and differences of force that operate in a field of struggle: it amounts to "taking a position in this field" (69), as Derrida puts it.[2]

Holding on to this understanding of the body of philosophy teachers, Derrida brings to light the double task of deconstruction. He explains that, by definition and necessity, his philosophical practice, which he calls deconstruction, is concerned with the contents as well as with the institution of philosophical teaching. "For a long time, therefore," Derrida writes, "it has been necessary . . . that deconstruction not limit itself to the conceptual content of philosophical pedagogy, but that it challenge the philosophical scene, all its institutional norms and forms" (72). This necessity is explained by Derrida a few pages later, when he weaves together the deconstruction of what he calls "phallogocentrism" and the deconstruction of the *universitas* as its institutional embodiment. From the perspective of my analyses, phallogocentrism converges with the tradition of the *logos spermatikos*—namely, with an organization of knowledge and of the relations among ontological regions that is based on the primacy of philosophy and, ultimately, on a certain understanding of genesis and life. Derrida observes:

> The deconstruction of phallogocentrism as the deconstruction of the onto-theological principle, of metaphysics, of the question "What is?," of the subordination of all fields of questioning to the onto-encyclopedic instance, and so

forth, such a deconstruction tackles the root of the *universitas* [university; totality]: the root of philosophy as teaching, the ultimate unity of the philosophical, of the philosophical discipline or the philosophical university as the basis of every university. The university is a philosophy. A university is always the construction of a philosophy. (2002b, 73)

But this does not mean that deconstruction brings about the end of philosophy and of the teaching of philosophy, their "death" (73), as Derrida suggests by referring to the Hegelian concepts of the beginning and the end of philosophy.[3] The implications of such an end would be contrary to the task of deconstruction. It would entail abandoning the field to other forces and interests that have already been solidary with phallogocentrism: forces "that have an interest in installing a properly metaphysical dogmatics—more alive than ever, in the service of forces that have from time immemorial been connected to phallogocentric hegemony" (73). Furthermore, deconstruction itself, as a teaching body, can neither be neutralized nor naturalized. Deconstruction is still to be identified as an "effect" (74) of the very field of forces that is under its systematic scrutiny. Within this framework, Derrida presents some of the proposals advanced by the GREPH in response to the Haby reform as the outcomes of the practical critique of the institutional body of philosophy and philosophy teachers that he calls "deconstruction." He writes:

> A rigorous and efficient deconstruction should at once develop the (practical) critique of the philosophical institution as it stands *and* undertake a positive, or rather affirmative, audacious, extensive and intensive transformation of a "philosophical" teaching. No longer a new *university design*, in the eschaton-teleological style of what was done under this name in the eighteenth and nineteenth centuries, but a completely other type of proposal, deriving from another logic and taking into account a maximum of new data of every kind, which I will not begin enumerate today. (74)

Derrida uncovers the articulation of power and the teaching body by taking up the exemplary case of his own position at the *Ecole Normale Supérieure* (ENS) as a teaching assistant of history of philosophy ("the position defined since the nineteenth century as that of d'*agrégé-répétiteur*,"

75). He recognizes in this position a power that is representative of what he designates as "the demand" or "power" that "dominates the system" (75). A few pages later, he explains that the "program" constitutes the knot of the representative relationship between these two powers. Implicitly referring to the Haby reform, produced without consulting the teaching body, Derrida observes that "a nonphilosophical and nonpedagogic power intervenes to determine who (and what) will determine, in a decisive and absolutely authoritarian fashion, the program and the filtering and coding mechanisms of all teaching" (78). Therefore, the very task of the deconstruction of the institutional body of philosophy coincides with the deconstruction of the programme and of the power attached to it. Derrida describes this task as follows:

> One of the difficulties in analyzing it stems from the fact that deconstruction must not, cannot only, choose between long and barely mobile networks and short and quickly outdated ones, but must display the strange logic by which, in philosophy at least, the multiple powers of the oldest machine can always be reinvested and exploited in a new situation. That is a difficulty, but it is also what makes a *quasi-systematic* deconstruction possible by protecting it against any empiricist light-headedness. These powers are not only logical, rhetorical, and didactic schemas. Nor are they even essentially philosophemes. They are also sociocultural or institutional operators, scenes or trajectories of energy, clashes of force that use all sorts of representatives. (79)

Once again, deconstruction does not aim to detach power from the teaching body but, rather, to single out the irreducible articulation of the two, at the same time as it inscribes itself within the field of struggle for mastery and domination.[4] In an incidental remark, Derrida traces this field back to the space of dissemination and genesis in general. "Wherever teaching takes place, therefore," he writes, "—and in the philosophical par excellence—there are, within that field, *powers*, representing forces in conflict, dominant or dominated forces, conflicts, contradictions (what I call *effects of differance*)" (79).[5] Given that the political consists precisely in taking sides in the aforementioned field of struggle, deconstruction cannot be interpreted as a destruction of the political or as a nonpolitical practice. Furthermore, the struggle in which the teaching body is engaged exceeds the confrontation between the philosophical and the nonphilosophical, so long as philosophy cannot be dissociated from the field of forces, interests, and powers. Derrida thus describes this field as *khōra*, as the anagrammatic layer of tropic movements, and thus as the structure or the structural law that allows us to think the historical totality of geneses. He writes:

> There could therefore never be *one* teaching body or *one* body of teaching . . . : one homogeneous, self-identical body suspending within it the oppositions (for example, the politics) that the place outside it, and sometimes defending PHILOSOPHY IN GENERAL against the aggression of the nonphilosophical from the outside. If there is a struggle regarding philosophy, then, it is bound to have its place inside as well as outside the philosophical "institution." (80)

In two other texts included in *Right to Philosophy I*, "The Crisis in the Teaching of Philosophy" (1978) and "Philosophy and its Classes" (1975), Derrida rejects the idea that the Haby reform is against philosophy. This reform takes place within the field of forces and interests and engages in the struggle for mastery and domination. In other words, it is engraved on *khōra*. Therefore, it must be philosophical. In the first of the two aforementioned texts, after recalling the key features of this reform (primacy of techno-scientific disciplines, destruction of philosophical criticism and promotion of noncritical humanities), Derrida explains that a more or less implicit philosophy undergirds it. "The Haby reform," he observes, "does not represent an antiphilosophy, but rather certain forces linked to a certain philosophical configuration, which, in a historico-political situation, have an interest in favoring this or that institutional structure" (111). In the second text, Derrida affirms that the aim of the Haby reform cannot be identified as the radical destruction of philosophy so long as the reform embodies the interest of a specific philosophical force (or difference of forces) in gaining mastery: "To impose an apparatus capable of inculcating a philosophy or maintaining a certain philosophical type, a philosophical force or group of forces, in a dominant position" (165). More precisely, the text of the reform, entitled *For a Modernization of the Education System*, is "a philosophical text" (165) committed to mastering the philosophical forces of critique and deconstruction.

II.

In these concluding paragraphs, I focus on the interpretation of Marx's genealogy of ideology—and thus of the institutional bodies that struggle for domination in a historico-political situation—which Derrida elaborates in the final lectures of the unedited seminar on *GREPH*. This interpretation highlights the relationship between the analyses of the teaching body and the general conditions of genesis (*khōra, différance,* etc.). Through a selective reading of the third *Economic and Philosophical Manuscript* (1844), Derrida points out that Marx explains the ideological process by linking it to the division of labor and its naturalization in the bourgeois economy. Furthermore, in the last lecture, he comments on a text from Marx's *German Ideology* (1845) that allows him to bring the genealogy of ideology beyond the limits of sociality and consciousness and to situate it on the very space of biological genesis. The text reads:

> With these [increased productivity and the increase of needs] there develops the division of labour, which was originally nothing but the division of labour in the sexual act, then the division of labour which develops spontaneously or "naturally" by virtue of natural predisposition (e.g., physical strength), needs, accidents, etc., etc. (Marx 1972, 51)

On Derrida's reading, here Marx admits that the division of labor and thus the process of ideology have been at work since sexual difference and thus within the space of biological and natural life. As I have demonstrated in the previous chapters, Derrida understands sexual difference as an irreducible trait of the mortal germ, in the wake of (and yet behind) Hegel's philosophy of nature. Therefore, through this excerpt from the *German Ideology*, he argues for the convergence between ideal and biological genesis and thus conceives of the mortal germ as the minimal structure of genesis in general. But Derrida also focuses on another element that he finds at play in the convergence emphasized in Marx's text—that is, the drive for domination and a certain effect of power. In so doing, I remark, he further develops his understanding of dissemination, thus providing the biological premises for his analysis of institutional bodies. He writes:

Linking [*lier*] the ideological process—before its very consciousness and effectivity—to sexual difference, this does not only entail that phantasmatization (*Einbildungskraft*) is not the classical type of conscious representation in progress towards the truth. It also amounts to linking imagination to a difference of forces within relations of domination [*domination*] and in a space of contradiction, or, in any case, of agonistic difference. In the case of what Marx calls natural DT [division of labor] (sexual DT, etc.), DT takes place in function of differences of forces [*la DT se fait en fonction des différences de forces*] (for instance, *Körperkraft*). As it is linked [*liée*] to a struggle that puts forces into play and it aims to mastery [*maîtrise*], ideality becomes, is itself a power [*pouvoir*] tending to domination (hence, the difficulty of demarcating the material from the spiritual: a technique is already an ideality). (Derrida 1974–1975, 9.9)[6]

It is worth recalling that this discovery of power and of the drive for mastery and domination is also at stake in a key passage of the lecture course entitled *La Vie la mort* (*Life-Death*), taught in 1975. Here, Derrida draws attention to the concept of *Bemächtigungstrieb* that Freud puts forth in *Beyond the Pleasure Principle* (1920), a concept that Derrida translates into "drive for mastery," "drive for power," and, more appropriately, "drive for domination" *(pulsion de maîtrise/puissance/emprise)*.[7] In particular, he focuses on the entry that, in *The Language of Psychoanalysis* (1967), Laplanche and Pontalis dedicate to the exploration of *Bemächtigungstrieb* throughout Freud's work.[8] This passage is further elaborated by Derrida in the later essay entitled "To Speculate—On 'Freud'" (published in *The Postcard: From Socrates to Freud and Beyond*, 1980), which consists in the revised version of the part of *La Vie la mort* devoted to the reading of Freud's *Beyond the Pleasure Principle*.[9] In this text, Derrida understands the drive for domination (*pulsion d'emprise, Bemächtigungstrieb*) as the minimal condition for the drive in general—that is, for its self-relation or ipseity.

NOTES

PREFACE

1. In the last chapter, I examine this paradigm as it is elaborated in the work of the French molecular biologist François Jacob, *The Logic of Life: History of Heredity* (1970).

2. For a description of this process, I refer to the article entitled "Noise as a Principle of Self-Organization" (1972) in Atlan 2011 (95–113). For an introduction to Atlan's work, see Geroulanos and Meyers's "Introduction to Complexity" (1–31).

3. On the relevance of this passage for French biological thought and philosophy, see the concluding pages of Georges Canguilhem's entry on "*Vie* [Life]" in the *Encyclopaedia Universalis* (Canguilhem 1989, 552–53).

4. For a seminal reading of Derrida's early engagement with the biology of his time, see Johnson 1993 (in particular, chapter 5, 142–87). Johnson demarcates Derrida's understanding of biological genesis from the paradigm of genetics and relates it to what he calls a "philosophy of evolution" (169). I refer the reader to chapter 5 for a more detailed discussion of Johnson's interpretation of Derrida's text.

5. *GREPH* is the acronym for *Group de recherche pour l'enseignement de la philosophie* (Research Group on the Teaching of Philosophy), which was officially set by Derrida and others in 1975.

INTRODUCTION

1. For another account of this struggle, see *Theodicy* (1710) section 201 in Leibniz 1952, 256:

> One may say that as soon as God has decreed to create something there is a struggle between all the possibles, all of them

laying claim to existence, and that those which, being united, produce most reality, most perfection, most significance carry the day. It is true that all this struggle can only be ideal, that is to say, it can only be a conflict of reasons in the most perfect understanding, which cannot fail to act in the most perfect way, and consequently to choose the best. Yet God is bound by a moral necessity, to make things in such a manner that there can be nothing better: otherwise not only would others have cause to criticize what he makes, but, more than that, he would not himself be satisfied with his work, he would blame himself for its imperfection; and that conflicts with the supreme felicity of the divine nature.

2. See also *Monadology* (1714) section 54 in Leibniz 1989, 648.

3. For another reading of this text, see Johnson 1993, 23–26, which, however, does not highlight the decisive break with Leibniz's notion of compossibility that Derrida's discovery of *sur-compossibilité* (overassemblage or pure equivocity) brings about. For Leibniz, a possible universe is made up by a collection of possible compossibles—that is, of the possibles that can exist together (cf. Leibniz 1989, 662).

4. For an analogous scene of responsibility, anguish, and decision, see how, almost thirty years later, Derrida (2002) describes the politico-juridical decision in "Force of Law":

It is a moment of suspense, this period of *epokhē*, without which there is, in fact, no possible deconstruction. It is not a simple moment: its possibility must remain structurally present to the exercise of all responsibility if such responsibility is never to abandon itself to dogmatic slumber, and therefore to deny itself. From then on, this moment overflows itself. It becomes all the more anguishing. But who will claim to be just by economizing on anguish? This anguishing moment of suspense also opens the interval of spacing in which transformations, even juridicopolitical revolutions, take place. (248–49)

5. Derrida quotes *Towards a Critique of Hegel's Philosophy* from Althusser's edition of Feuerbach's writings, entitled *Manifestes philosophiques* (1960). I come back to this text several times, as Derrida refers to it in key moments of the writings that are examined here.

6. For the use of *Aufhebung* in the examined section of the *Encyclopedia Philosophy of Nature*, see Hegel 1970, 414: "The genus preserves itself only through the destruction of the individuals who, in the process of genera-

tion, fulfill their destiny and, in so far as they have no higher destiny, in this process meet their death."

7. As Althusser points out in his introduction to the French edition of Feuerbach's text: "But *Gattung* is not merely a biological category. It is essentially a theoretical and practical category. In the privileged case of humans, it appears at once as the true transcendental horizon that makes the constitution of all theoretical significations possible and as the practical Idea that gives history its sense" (Feuerbach 1960, 14 [my translation]). For a further exploration of the theory of *Gattung* in the later essay "On Feuerbach" (1967), see also Althusser 2003, 137–50.

8. Derrida addresses the Hegelian source in an explicit fashion only later, in the final pages of "Freud and the Scene of Writing" (1966), in which he dissociates the trace, as a natural and biological seed, as a mortal germ that bears within itself the death of its genitors, from the incorruptible substance of the son of God, in which the father circulates himself and thus is fully present. See Derrida 1978, 289: "The trace is the erasure of selfhood [*effacement de soi*], of one's own presence [*de sa propre présence*], and is constituted by the threat or anguish of its irremediable disappearance, of the disappearance of its disappearance. An inerasable [*ineffaceable*] trace is not a trace, it is a full presence, an immobile and incorruptible substance, a son of God, a sign of *parousia* and not a seed, that is, a mortal germ [*germe de mort*]." For a later elaboration of the scene of generation as a configuration of angustia, inscription and alienation (without return), I recall the following note in *Glas* (1974):

> Another form of the same question: can a family name be translated? Strangulation: the singularity of the general, the classification of the unique, the tightening structure of a grip in which the concept conceives, limits and delimits itself [*L'étranglement: la singularité du général, la classification de l'unique, la structure rétrécissante d'une étreinte où se conçoit, limite et délimite le concept*]. . . . The passion of the proper name: never to let itself be translated—according to its desire—but to suffer translation—which is intolerable to it. (Derrida 1986, 20)

For a closer reading of this text, I refer to chapter 4.

9. A page later, Derrida accounts for the tragedy of the book as the minimal condition for genesis and life: "On what could books in general *live*, what would they be if they were not alone, so alone, infinite, isolated worlds?" (1978, 11).

10. For the Leibnizian notion of expression, see *Monadology* section 55: "Now this mutual connection or accommodation of all created things to each other and of each to all the rest causes each simple substance to have

relations which express all the others and consequently to be a perpetual living mirror of the universe" (Leibniz 1989, 648). I observe that, for Derrida, writing *qua* engendering disrupts the relation of *resemblance/rassemblence* (resembling/reassembling) between inscription/seed and sense/species. On this point, see Derrida 1978, 8–9: "To write is not only to know that through writing, through the extremities of style, the best will not necessarily transpire, as Leibniz thought it did in divine creation, nor will the transition to what transpires always be *willful*, nor will that which is noted down always infinitely *express* the universe, resembling and reassembling it."

11. On the relevance of this legacy for Derrida's work, Geoffrey Bennington has an important note:

> Writing constrains meaning . . . in the anguished passage which both limits and liberates its "potential of pure equivocality," and does so by leaving a trace that is not exhausted in its immediate context, but emancipates itself and opens itself to the risk and the chance of a pure transmission that never will be pure, but which, through this very impurity, opens the concrete possibility of reading itself. (We'd need to reread here Derrida's Introduction to Husserl's *Origin of Geometry* the better to follow the detail of this logic according to which a certain ideality liberated by writing is then reinscribed in a certain mundane facticity which alone, however, "delivers the transcendental" (71). (Bennington 2010, 124)

12. For a close reading of these sections, see Lawlor 2002, 116–25. Here I am interested in highlighting the constitutive role of writing that Derrida interprets as the legacy of the *Origin* and that links writing to engendering in general.

13. From this paradoxical evidence Derrida concludes that the relationship between language and sense as well as the one between the factual and the transcendental need to be reformulated: "The paradox is that, without the apparent fall back into language and thereby into history, a fall which would alienate the ideal purity of sense, sense would remain an empirical formation imprisoned as fact in a psychological subjectivity—in the inventor's head. Historical incarnation sets free the transcendental, instead of binding it" (Derrida 1989, 78).

14. Cf. Derrida 1989, 87–89.

15. On this section, see Lawlor 2002, 106–07. For a further elaboration of this comparative reading, see the section of *Grammatology* in which Derrida develops "a new transcendental aesthetics" that breaks with the Kantian as well as Husserlian legacy (cf. Derrida 1974, 290–91).

16. In a note, Derrida acknowledges the convergence between his comparative reading and the analysis that Eugen Fink elaborates in "The Phenom-

enological Philosophy of Edmund Husserl and Contemporary Criticism" (under Husserl's approval). For a preliminary attempt to situate Derrida's reading in light of the legacy of Kant in French epistemology, see Baring 2011, 113–45.

17. For the translation of *Leistung* by "production," see the related footnote in Derrida 1989, 40:

> Among all the translations already proposed for the notion of *Leistung*, so frequently utilized in the *Origin*, the word 'production' seemed to overlay most properly all the significations that Husserl recognizes in this act that he also designates by some complementary notions: *pro-duction*, which leads to the light, constitutes the 'over against us' of Objectivity; but this bringing to light is also, like all production (*Erzeugung*) in general, a creation (*Schöpfung*) and an act of formation (*Bildung*, *Gestaltung*), from which comes ideal objectivity as *Gebilde*, *Gestalt*, *Erzeugnis*, and so on.

18. On this point, see Hegel 1999, 178 and Derrida 1986, 106.

19. Derrida had already played with the term *de-funct* in his *Introduction* to Husserl's *Origin of Geometry* (section 7), where the term accounts for the detachment of a text from any relationship with a transcendental subject and thus confines the text itself to a prehistorical layer. See Derrida 1989, 88:

> If the text does not announce its own pure dependence on a writer or reader in general (i.e., if it is not haunted by a virtual intentionality), and if there is no purely juridical possibility of it being intelligible for a transcendental subject in general, then there is no more in the vacuity of its soul than a chaotic literalness or the sensible opacity of a defunct designation, a designation deprived of its transcendental function. The silence of prehistoric arcana and buried civilizations, the entombment of lost intentions and guarded secrets, and the illegibility of the lapidary inscription disclose the transcendental sense of death as what unites these things to the absolute privilege of intentionality in the very instance of its essential juridical failure [*en ce qui l'unit a l'absolu du droit intentionnel dans l'instance même de son échec*].

On this passage, see also Marrati 2005, 35–39.

20. See the section "Programme" in part I of *Of Grammatology* (Derrida 1974, 9–10). I come back to this point throughout the book and, in particular, in the last chapter.

21. On this point, see Vitale 2014, 99.

22. On the articulation of geneticism and genetics, I refer to Derrida 1981, 50, and to the reading of this text developed in chapter 5.

23. For the description of the phenomenology of spirit (and of the pure concept that engenders its own actuality) in the preface to *Phenomenology*, see Hegel 1977, 14:

> That the True is actual only as system, or that Substance is essentially Subject, is expressed in the representation of the Absolute as *Spirit*—*the* most sublime Notion and the one which belongs to the modern age and its religion. The spiritual alone is the *actual;* it is essence, or that which has *being in itself;* it is that which *relates itself to itself* and is *determinate,* it is *other-being* and *being-for-self,* and in this determinateness, or in its self-externality, abides within itself; in other words, it is *in and for itself.* But this being-in-and-for-itself is at first only for us, or *in itself,* it is spiritual *Substance.* It must also be this *for itself,* it must be the knowledge of the spiritual, and the knowledge of itself as Spirit, i.e. it must be an *object* to itself, but just as immediately a sublated object, reflected into itself. It is for *itself* only for *us,* insofar as its spiritual content is generated by itself. But insofar as it is also for itself for its own self, this self-generation, the pure Notion, is for it the objective element in which it has its existence, and it is in this way, in its existence for itself, an object reflected into itself. The Spirit that, so developed, knows itself as Spirit, is *Science;* Science is its actuality and the realm which it builds for itself in its own element.

24. For this reading of the phenomenological development as well as of its necessity and endpoint, see Hyppolite 1974, 11–26.
25. Cf. Derrida 1978, 26.
26. Cf. Roger 1999, 266
27. For a crucial elaboration of this point, I refer to Derrida's later essay "*Ousia* and *Grammē*" (1968; in particular, Derrida 1982, 62). See also John Protevi's analysis in "The Circle and the Trace: '*Ousia* and *Grammē*'" (Protevi 1994, 76–110).
28. In *Metaphysics* 9.1048b35–1049a18, Aristotle illustrates the concept of potentiality by recalling the example of the human seed:

> E.g., is *earth* potentially a man? No—but rather when it has already become *seed,* and perhaps not even then, as not everything can be healed by the medical art or by chance, but there is a certain kind of thing which is capable of it, and only this is potentially healthy. . . . The seed is not yet potentially a man, for it must further undergo a change in a foreign medium. But when through

its own motive principle it has already got such and such attributes, in this state it is already potentially a man; while in the former state it needs another principle, just as earth is not yet potentially a statue, for it must change in order to become bronze.

CHAPTER 1

1. On Heidegger's elaboration of destruction against Hegel's movement of refutation, see session 1 of the lecture course. For a preliminary analysis of Derrida's comparative reading of Hegel and Heidegger in the course, see Gratton 2015.
2. For Derrida's reference to this passage, see Derrida 2016, 26.
3. On this point, see Derrida 2016, 26–27.
4. Cf. Derrida 2016, 30–31.
5. See Derrida 2016, 31: "'Telling stories,' then—that is, giving oneself over to a mythological discourse (I'm finally arriving at the reference I announced)—is something one tried to renounce for the first time in philosophy precisely at the moment when the problem of being announced itself as such. Heidegger does not multiply references; he merely cites this 'first time' when storytelling was dismissed in the face of the problem of being. This is Plato's *Sophist* (242e)."
6. For a deeper engagement with the structure of the *Timaeus* and the reasons why Derrida prefers this dialogue (versus Heidegger's *Sophist*), I refer to the analysis of "Plato's Pharmacy" below and to the reading of *khōra* that I propose in the next chapter.
7. For a close reading of Timaeus's discourse, in line with Derrida's observations, see Sallis 1999, 52–55.
8. On the use of "necessity," see Derrida 2016, 36.
9. Cf. Derrida 1981, 73–75. On the relevance of Plato's exclusion of writing (as it is analyzed in "Plato's Pharmacy") for Derrida's subsequent work, see Naas 2014, 237–46.
10. See Derrida 1981, 74: "And at the same time, through writing or through myth, the genealogical break and the estrangement from the origin are sounded"; and 77: "The specificity of writing would thus be intimately bound to the absence of the father. Such an absence can of course exist along very diverse modalities, distinctly or confusedly, successively or simultaneously."
11. Cf. Derrida 1981, 96–97. For the determinations of "father" and "noble birth," see *Phaedrus* 257b and 261a.
12. Cf. *Phaedrus* 264b–c, also quoted in Derrida 1981, 79: "But I think you would assert this, at any rate: that every speech, just like an animal, must be put together to have a certain body of its own, so as to be neither headless

nor footless but to have middle parts and end parts, written suitably to each other and to the whole."

13. In particular, I think of *Glas*, which Derrida himself announces as a work "on Hegel's family and on sexual difference in the dialectical speculative economy" (1982, 76). For a reading of *Glas* from the perspective on genesis developed here, I refer to chapter 4.

14. For Derrida's first definition of the *logos spermatikos*, see my introduction. Here I quote again the passage commented on above from Derrida 1978, 26–27: "But when one is concerned with an art that does not imitate nature, when the artist is a man, and when it is consciousness that engenders, preformationism no longer makes us smile. *Logos spermatikos* is in its proper element, is no longer an export, for it is an anthropomorphic concept."

15. For an interpretation of these pages dedicated to the paternal thesis, see also the reading proposed by Michael Naas in his seminal "Earmarks: Derrida's Reinvention of Philosophical Writing in 'Plato's Pharmacy.'" This chapter may also be read as an elaboration of Naas's provocation. He interprets the father-logos relation from the perspective of the anagrammatic structure of Plato's text, which is uncovered later in my text. Therefore,

> paternity . . . the father, is himself always the effect of discourse, of logos, the very thing that is presented in Plato as being the son or off-spring of a living, speaking father. The expression "father of logos" is thus not a simple metaphor, Derrida argues, for if "the father is always father to a speaking/living being . . . it is precisely logos that enables us to perceive and investigate something like paternity." In other words, it would be from the son, from logos, from the supplement, from the position of writing as a system of differences, that the father would be able to be called a father, that the father would come to be identified as a father. What makes the inscription of the father possible in a system is, in a word, his son, his supplement, the one who comes after him, who thus threatens not just to supplement him but to supplant him, "the father of the logos" is not a metaphor insofar as the father is the effect of the logos and thus it can be conceived of only from the logos. In other words, the expression already implies the alienation of origin that determines writing. (Naas 2010, 49–50)

16. See Derrida 1981, 86:

What then, are the pertinent traits for someone who is trying to reconstitute the structural resemblance between the Platonic and the other mythological figures of the origin of writing? The bringing out of these traits should not merely serve to determine each of the significations within the play of thematic oppositions as they have been listed here, whether in Plato's discourse or in a general configuration of mythologies. It must open onto the general problematic of the relations between the themes and the philosophemes that lie at the origin of western *logos*. That is to say, of a history—or rather, of History—which has been produced in its entirety in the *philosophical* difference between *mythos* and *logos*, blindly sinking down into that difference as the natural obviousness of its own element.

17. Cf. Derrida 1981, 88–90.
18. Cf. Derrida 1981, 96.
19. For the description of this other pole, see Derrida 1981, 97: "The magic virtues of a force whose effects are hard to master, a dynamics that constantly surprises the one who tries to manipulate it as master and as subject."
20. Derrida's pages on Levinas's notion of absolute exteriority, in "Violence and Metaphysis: Essay on Levinas's *Totality and Infinity*," may be read as a protocol on the practice of erasing (*effacer*) as well as on the impossibility of erasing syntax and the consequences of this impossibility. Cf. Derrida 1978, 139–40.
21. On the anagrammatic displacement through which, according to Rousseau and Saussure, writing (grapheme) has usurped language (phoneme) from the outset, see Derrida 1974, 37ff.
22. See, for instance, Derrida 1978, 141–42, on Hegel's account of the delight of speculative thought before the existence of words with speculative meaning in the German natural language. In chapters 3 and 4, I undertake an overall analysis of Derrida's reading of the Hegelian solution of the equivocity between natural and philosophical language.
23. For an explicit treatment of these syntactical and tropic movements as the vigil of philosophy, I refer to Derrida's essay "White Mythology. Metaphor in the Text of Philosophy" (1972). In particular, see Derrida, 1982, 228–29:

> The constitution of the fundamental oppositions of the metaphorology (physis/techne, physis/nomos, sensible/ intelligible, space/ time, signifier/signified, etc.) has occurred by means of the history of a metaphorical language, or rather by means of "tropic" movements which, no longer capable of being called by a philosophical name—i.e. metaphors—nevertheless, and for the same reason, do not make up a "proper" language. It is from beyond the difference between the proper and the nonproper that the effects of propriety

and nonpropriety have to be accounted for. By definition, thus, there is no properly philosophical category to qualify a certain number of tropes that have conditioned the so-called "fundamental," "structuring," "original" philosophical oppositions.

For a recent reading of this text, see Gasché 2014. In this chapter, I show that tropic movements account not so much for the shared field of rhetoric and philosophy (as Gasché suggests) as for the element of metaphorical exchanges among regional discourses and, more generally, for linguistic as well as zoological concatenations or geneses.

24. Plato's text reads: "The triangles in us are originally framed with the power to last for a certain time beyond which no man can prolong his life" (Derrida 1981, 101).

25. On the meaning of *pharmakeus*, see Derrida 1981, 117–19. In chapter 6, he writes:

> The circuit we are proposing is, moreover, all the more legitimate and easy since it leads to a word that can, on one of its faces, be considered the synonym, almost the homonym, of a word Plato "actually" used. The word in question is *pharmakos* (wizard, magician, poisoner), a synonym of *pharmakeus* (which Plato uses), but with the unique feature of having been overdetermined, overlaid by Greek culture with another function. Another role, and a formidable one. The character of the *pharmakos* has been compared to a scapegoat. (130)

26. The sources of this chapter of "Plato's Pharmacy" are summarized in an extended footnote in Derrida 1981, 130–32.

27. Cf. Derrida 1981, 131–33.

28. Derrida highlights the constructed character of the *pharmakos* and of the date of purification as requirements for the mastery of the critical instance. Cf. Derrida 1981, 133: "These exclusions took place at critical moments (drought, plague, famine). *Decision* was then repeated. But the mastery of the critical instance requires that surprise be prepared for: by rules, by law, by the regularity of repetition, by fixing the date."

29. For a preliminary analysis of this relationship, see Vitale, 2013.

30. For the analysis of these passages, see Derrida 1981, 148–49.

31. On the "irreducible analogy" that Derrida highlights in this passage, see Naas 2010, 46: "Though the Platonic text will always try to convince us that these borrowings are 'mere' analogies, we are now prepared to see that they just might well be the irreducible analogies—or anagrams—that at once impose and undercut the Platonic system."

32. See Derrida 1981, 153:

Writing and speech have thus become two different species, or values, of the trace. One, writing, is a lost trace, a nonviable seed, everything in sperm that overflows wastefully, a force wandering outside the domain of life, incapable of engendering anything, of picking itself up, of regenerating itself. On the opposite side, living speech makes its capital bear fruit and does not divert its seminal potency toward indulgence in pleasures without paternity.

33. See Derrida 1981, 165: "As far as the rules of concordance and discordance, of union and exclusion among different things are concerned, this *sumplokē* 'might be said to be in the same case with the letters of the alphabet' (*Sophist* 253a; cf. also Plato's *Statesman* where the 'paradigm' of the *sumplokē* is equally *literal*, 278a–b)."

34. For Plato's text, see *Sophist* 253b–c: "And as classes are admitted by us in like manner to be some of them capable and others incapable of intermixture, must not he who would rightly show what kinds will unite and what will not, proceed by the help of science in the path of argument? And will he not ask if the connecting links are universal, and so capable of intermixture with all things; and again, in divisions, whether there are no other universal classes, which make them possible?"

35. For Derrida's deconstruction of the classical notion of the *symploke* (as Plato elaborates it in the *Statesman*), see Gasché 1986, 93–99.

CHAPTER 2

1. It is worth remarking that in the texts that I examine in this chapter, Derrida highlights the difference between Timaeus's recourse to a third term and the one performed by the Stranger in Plato's *Sophist* (259e). As he had pointed out in the 1964–1965 lecture course, in *Being and Time*—think of the exergue as well as of Introduction section 2—Heidegger interprets the Stranger's introduction of the third term of Being as the first step taken by philosophy onto the question of Being, beyond mythology. Now, Derrida understands Plato's reference to *khōra*, in the *Timaeus*, as a step taken beyond philosophy and thus beyond the *Sophist*, including the *Sophist* of Heidegger's *Being and Time*. An implicit confrontation with Heidegger's preference for the *Sophist* is thus at stake, here, in Derrida's opting for the *Timaeus*.

2. Cf. Derrida 1985–1986, "Questions. Le 11 décembre 1985," 10.

3. On the methodological presuppositions of this investigation, see Loraux 1993, 14–15.

NOTES TO CHAPTER 2

4. On Aspasia's praise for the fatherland as the true mother of the Athenians, see Derrida 1985–1986, "CHORA (suite)," 9. It is worth recalling that Derrida goes back to this text from the *Menexenus* in *The Politics of Friendship* (1994), when he highlights the link between equality and nobleness of birth (namely, autochthony and *eugeneia*), on the one hand, and fraternal aristo-democracy, on the other. On this point, see Derrida 1997, 92–96. In this text, apropos of the necessary precautions that the reading of Aspasia's discourse requires, Derrida observes: "The question whether this staging is ironic (we shall return to this point), whether the most common logic and rhetoric, the most accredited eloquence of *epitaphios*, is reproduced by Plato in order to belittle it, only gives that much more sense to the fictive contents of the discourse attributed to Aspasia, that courtesan who, moreover, plagiarizes another funeral oration and mouths once again the 'fragments' of a discourse by Pericles (236b)" (92).

5. Cf. Derrida 1985–1986, "CHORA (suite)," 11–12.

6. For a reconstruction of the myth of Erichthonius, see Loraux 1981, 7–8 and 32–33.

7. On the question of generations, Loraux writes that, through autochthony, orators affirm "the permanence of the Athenian principle through successive generations, which, linked one to the next, ensure the continuation of the Same'" (1981, 51).

8. Cf. Derrida 1970–1971, 1.10, where *khōra* is determined as "the absolute outside" of the philosophical couple.

9. For a close reading of this text, I refer to Sallis 1999, 91–97. The author marks this other beginning as the point of departure of what he calls "chorology."

10. This discussion comes back in the 1985–1986 texts on *khōra* (cf. Derrida 1985–1986, "22 janvier," 19) and in the concluding pages of *Khōra*.

11. Plato's text reads:

> For this fresh start of ours, we need to take account of more than we did before. Earlier we distinguished two types of things, but now we have to disclose the existence of a third kind, different from the others. Our earlier discussion required no more than the two—the model, as we suggested, and the copy of the model, the first being intelligible and ever consistent, the second visible and subject to creation—and we didn't distinguish a third at the time, on the grounds that these two would be sufficient. But now the argument seems to demand that our account should try to clarify this difficult and obscure kind of thing. (*Timaeus* 48e–49b)

12. In an important note on the obscurity ("meaninglessness") of *khōra*, Sallis observes:

Regarding what he calls "the thought of the χώρα," Derrida writes: "It would no longer belong to the horizon of meaning [*sens*], nor to that of meaning as the meaning of being" (*Khōra*, 22f.). Presumably, Derrida is alluding here to Heidegger and suggesting that the thought of the χώρα exceeds the horizon of Heidegger's thinking, at least of the inscription of that thinking that is governed by the question of the meaning of being. In Heidegger's own brief discussions of the χώρα, which conflate χώρα and τόπος and link Platonism to the transformation of the essence of place into space defined as extension, there is little to suggest any originary engagement with the Platonic discourse on the χώρα. (1999, 111)

13. See Derrida 1970–1971, 4.8: "If it is true that the transition to the third in the *Timaeus* . . . means . . . passing to a third that has not been *anticipated* [my emphasis], then this movement that opens onto *khōra* is irreducible to the whole ontological hermeneutics of explanatory nature [*toute l'herméneutique ontologique de style explicitant*] that goes from the *Timaeus* to *Sein und Zeit*."

14. See Derrida 1970–1971, 4.9: "This is why this third is not there in order to give place to a dialectic, a dialectical philosophy . . . the transition to *khōra* as a third seems to suspend the order and security of dialectical and philosophical knowledge, it is the moment when we leave the epistemic security that the presence of the meaning *being* [*du sens* être] assures in language."

15. Cf. *Timaeus* 49b–50c. In particular, I refer to the analogy with gold, which is still a thing that enters *khōra* as its receptacle:

> I'd better go back over what I've been saying and try to make it even clearer. Imagine someone who moulds out of gold all the shapes there are, but never stops remoulding each form and changing it into another. If you point at one of the shapes and ask him what it is, by far the safest reply, so far as truth is concerned, is for him to say "gold"; he should never say that it's "a triangle" or any of the other shapes he's in the process of making, because that would imply that these shapes are what they are, when in fact they're changing even while they're being identified. However, he'd be content if you were, after all, also prepared to accept, with some degree of assurance, the reply "something of this sort." By the same argument, the same term should always be used in speaking of the receptacle of all material bodies, because it never is anything other than what it is: it only ever acts as the receptacle for everything, and it never comes to resemble in any

way whatsoever any of the things that enter it. Its nature is to act as the stuff from which everything is moulded—to be modified and altered by the things that enter it, with the result that it appears different at different times. (50a–c).

On this text, see Sallis 1999, 108–13.

16. See *Timaeus* 50e: "Think, for instance, of perfumery, where artisans do exactly the same, as the first stage of the manufacturing process: they make the liquids which are to receive the scents as odourless as possible. Or think of those whose work involves taking impressions of shapes in soft materials: they allow no shape at all to remain noticeable, and they begin their work only once they've made their base stuff as uniform and smooth as possible."

17. The text reads: "The *khōra* of the *Timaeus*, which is not at all the country, the land, the region in which we bury and were born. The autochthonous were born in their *khōra*. They bury soldiers in their *khōra*. Oedipus walks toward his land, his *khōra*. However, between this *khōra* and the one of the *Timaeus*, which is the unique and universal receptacle of what is in the cosmos, there is an absolute rupture, a sensible, marked heterogeneity" (Derrida 1985–1986, "22 janvier," 1–2).

18. As we see below, the power to receive, which seems to determine *khōra* ("she can"), is a strange power, a certain *dynamis*.

19. On the necessity of the *emphanisis*, see Derrida 1985–1986, "22 janvier," 30: "This necessity that we may easily consider hyper- or metaphilosophical cannot in any case be homogeneous to the dominant regime of the philosophical."

20. On this point, cf. Derrida 1985–1986, "22 janvier," 11. Here I understand *khōra*'s emphasized power of reception ("can") in light of the subsequent explanation of its *dynamis*.

21. On the interpretation of the *dynamis* of *khōra*, see Derrida 1985–1986, "22 janvier," 15: "If therefore *khōra* never leaves what the text calls her *dynamis*, her property, her potency, her capacity, this is because she remains always the same indetermination open to everything that can come to her. This *dynamis* has not the meaning of a virtuality that waits for passing into actuality, it is not a potential matter but, rather, a *dynamis* that remains *dynamis* independently from what comes and thus without being deprived of anything."

22. As the exergue of this chapter testifies, *khōra* keeps on playing a key role in Derrida's late texts such as *Rogues. Two Essays on Reason* (2003).

23. Cf. Derrida 1995, 89–91.

24. Cf. Derrida 1985–1986, "Questions. 8 janvier 1986," 2.

25. Alongside the word "structure," Derrida also has recourse to the terms "sum" and "process." See, for instance, Derrida 1995, 99: "She 'is' nothing other than the sum or process of what has just been inscribed 'on' her."

26. For a critical reading of Derrida's understanding of *mise en abyme*, see Gasché's essay "Structural Infinity" in Gasché 1994, 129–49.

27. Cf. Derrida 1970–1971, 3.10 and Derrida 1985–1986, "22 janvier," 19.

CHAPTER 3

1. Cf. Derrida 1969–1970, 1.2–3. For the text of Aristotle's *Metaphysics* that Derrida implicitly refers to in his lecture course, see *Metaphysics* 6.1026a33:

> One might indeed raise the question whether first philosophy is universal, or deals with one genus, i.e. some one kind of being; for not even the mathematical sciences are all alike in this respect— geometry and astronomy deal with a certain particular kind of thing, while universal mathematics applies alike to all. We answer that if there is no substance other than those which are formed by nature, natural science will be the first science; but if there is an immovable substance, the science of this must be prior and must be first philosophy, and universal in this way, because it is first. And it will belong to this to consider being qua being—both what it is and the attributes which belong to it qua being.

2. In the next chapters we see that the link between family and the philosophical discourse constitutes also a guiding thread of Derrida's later analyses of the Hegelian text.

3. For the implications that, according to Derrida, the questioning of the primacy and universality of philosophical discourse bears for the understanding of the biology and the life sciences of his time, I refer to the last chapter of this book.

4. See Derrida 1969–1970, 1.5: "You know that if philosophy consists in refusing to credit a stranger procedure without appropriating it, for example a procedure belonging to a determinate science, this critical resistance has been exercised with a certain harshness and obstinacy, remarkable for their constancy, against mathematics and formal logic in general. However, this resistance has been commensurate with fascination [*a été à la mesure de la fascination*]." For the concept of *one's making oneself afraid*, I refer to chapter 4 of *Specters of Marx* (cf. Derrida 1994, 118–55).

5. In this chapter and the following ones, I quote the Hegelian text commented by Derrida in the English translation and emphasize the moments of Derrida's translation that are relevant to our reading.

6. This explains why natural language can be more speculative than the scientific one. On this point, see Derrida 1969–1970, 1.10.

7. On the (natural-)speculative word *Aufhebung*, see Nancy 2001, 15–16, where the author explains that, according to Hegel, *Aufhebung* is legible only through the presupposition of *Aufhebung* itself; and 60–62, in which he analyzes the preface to the *Science of Logic* also commented by Derrida. More generally, for Nancy's analysis of Hegel's philosophy of language, see chapter 3, entitled "Speculative Words" (Nancy 2001, 51–72). Finally, on the French translation and interpretation of *Aufhebung*, see the entry on *Aufhebung* in Cassin 2014.

8. See Derrida's remarks on this passage in Derrida 1969–1970, 1.11: "Hegel proposes the principle or concept of a general organization and history of poetic, scientific, philosophic, and so forth, language and discourse. The concept of culture—precisely, of *Bildung*—responds to this intention."

9. In the final page of the later essay "White Mythology," Derrida writes in the wake of his examination of Aristotle's philosophy of rhetoric: "a homonymy in which Aristotle recognized—in the guise of the Sophist at this point—the very figure of that which doubles and endangers philosophy" (Derrida 1982, 271).

10. Cf. Derrida 1978, 139–41.

11. In *New Essays on Human Understanding* (1690) section III.9, Leibniz explains that "the communicative use of words is also of two sorts, *civil* and *philosophical* [my emphasis]. The civil use consists in the conversation and practice of civil life. The philosophical use of words [is] such a use of them, as may serve to convey . . . precise notions, and to express certain truths in general propositions" (1996, 335).

12. On this point, I refer to the introduction to *Speech and Phenomena* (1967, see, in particular, Derrida 1973, 6–15) and to session 1 of the unedited 1964–1965 lecture course *La théorie de la signification dans les* Recherches logiques *et dans* Ideen I (*The Theory of Signification in the* Logical Investigations *and in* Ideas I), which constitutes the draft of the aforementioned published text.

13. For the text commented by Derrida, see Hegel 2010, 735:

> The absolute idea, as the rational concept that in its reality only rejoins itself, is by virtue of this immediacy of its objective identity, on the one hand, a turning back to life; on the other hand, it has equally sublated this form of its immediacy and harbors the most extreme opposition within. The concept is not only soul, but free subjective concept that exists for itself and therefore has personality—the practical objective concept that is determined in and for itself and is as person impenetrable, atomic subjectivity—but which is not, just the same, exclusive singularity; it is rather explicitly universality and cognition, and in its other has its own objectivity for its subject matter. All the rest is error, confusion,

opinion, striving, arbitrariness, and transitoriness; the absolute idea alone is being, imperishable life, self-knowing truth, and is all truth.

Derrida adds the following remark to the quotation of the Hegelian text: "We are thus before an equivocal signification, an exemplary one, that functions simultaneously in the scientific field where we have to do with natural and objective life and in the philosophical field where the spiritual life of the idea is at stake. Life is not only natural life as it lets itself be sublated, but is already sublated life [*vie déjà relevée*], as the life of the spirit, and sublating life [*vie relevante*], as the spiritual process of *Aufhebung*, for it is the life of the Idea that sublates itself" (1969–1970, 1.14).

14. The text continues as follows: "As treated in the philosophy of nature, as the life of nature and to that extent exposed to *the externality of existence*, life is *conditioned* by inorganic nature and its moments as idea are a manifold of actual shapes. Life in the idea is without such *presuppositions*, which are in shapes of actuality; its presupposition is the *concept* as we have considered it, on the one hand as subjective, and on the other hand as objective" (Hegel 2010, 677).

15. For Derrida's exploration of Hegel's understanding of the processes of natural life as images or metaphors of the processes of spiritual life, I refer to the analysis of *Glas* undertaken in the next chapter.

16. I emphasize Derrida's interest in the fact that Hegel takes recourse to the natural concept of force in order to account for a spiritual phenomenon.

17. For Hegel's elaboration of the transition from consciousness to self-consciousness, see Hegel 1977, 104–19.

18. On the determination "natural-speculative," see how Derrida explains, in *Glas*, the Hegelian concept of Christianity:

> Thus Christianity offers the example of a *naturally speculative* religion. Philosophy—speculative dialectics—will have been the truth of this religious representation of the speculative. *Just as German, the naturally speculative tongue in certain of its traits, relieves itself by itself in order to become the universal tongue* [my emphasis], so a historically determinate religion becomes absolute religion, and an absolute religion relieves its character of representation (*Vorstellung*) in order to become absolute truth. (Derrida 1986, 32)

19. For Derrida's full argument, see Derrida 1982, 219–20. For an analysis of this argument and of its consequences, see also Bennington 2014, 91–92, and Gasché 2015, 14–16.

20. For the lexicon of my hypothesis, I refer the reader to the aforementioned passage from "Plato's Pharmacy" in Derrida 1981, 99, as well as to the remarks that I develop in chapter 1.

21. For a close reading of this double destruction, see Gasché 2014, 9–10.

22. Bennington (2014) describes the generalization or the second death of the metaphor as metaphoricity:

> Once metaphor is generalized or disseminated in this way and has no contrary, no contrastive "proper," it can no longer strictly be called "metaphor." . . . This generalized or "originary metaphoricity," here, as in "The Double Session," associated by Derrida with the notion of a syntax in excess of any semantics (DIS, 193, 211, 220), then becomes, in a fashion entirely characteristic of Derrida's earlier work, one name for the general *milieu* out of which the classical opposition of the metaphorical and the proper could conceivably have emerged in the first place. The "proper" (and its associated values) is then no longer the primary term, but a secondary determination of this originary metaphoricity, and therefore never entirely proper at all. By this means we have avoided the symmetrical positions identified earlier and made some progress in what looked like an aporia: positing this "originary metaphoricity" endorses neither the metaphysical position nor its claimed 'poetic' or 'rhetorical' reduction. (96)

As I have observed apropos of Gasché's reading of "White Mythology," I push this understanding of the *milieu* to its limits by suggesting that what remains of the metaphor, the trace-seed, constitutes the genetic structure of language as well as of the living.

23. For a close reading of this double destruction, see Gasché 2014, 9–10.

CHAPTER 4

1. This chapter is a revised version of Senatore 2105, which was written as a response to Vitale 2015. I am indebted to Vitale's article for raising some of the issues that are re-elaborated here from the perspective of this book.

2. For Derrida's schematization of the nature-spirit relation, see the following remarks on the progress of *Sittlichkeit* through its determinations, as Hegel describes it in the *Philosophy of Right* (1821): "But as every sally of the spirit outside of itself has the general form of its other, to wit, nature; nature is the spirit outside of itself but also a moment of the spirit's return to self, so *Sittlichkeit* will entail this naturalness. That will be a spirit-nature. Its naturalness will resolve itself, reabsorb itself, spiritualize itself in proportion as *Sittlichkeit* will develop itself through the form of its moments, will exhaust the inner negativity that works (over) it, will produce itself by denying itself

as nature" (Derrida 1986, 15). Furthermore, on the "difference" implicit in the concept of "analogy," see Derrida 1986, 104: "In the analogy, the difference [between natural and spiritual life, metaphorical value and semantic tenor, etc.] remains essential."

3. For the Hegelian text examined by Derrida, see Hegel 1999, 155:

> For although, in the living shape or organic totality of ethical life, what constitutes the real aspect of that life is [to be found] in the finite, and therefore cannot in and for itself fully incorporate [*aufnehmen*] its own bodily essence into the divinity of that life, it nevertheless already expresses the absolute Idea of ethical life, albeit in a distorted form. Admittedly, ethical life does not inwardly unite into absolute infinity within itself those moments of the Idea which are of necessity kept apart; on the contrary, it has this unity only as a simulated negative independence, namely as freedom of the individual. But this real essence is nevertheless completely bound up with the absolutely indifferent nature and shape of ethical life; and if it must perceive this nature only as something alien, it does nevertheless perceive it and is at one with it in spirit.

4. For an analogous interpretation of the Hegelian philosophy of nature, see Althusser's doctoral thesis entitled "On Content in the Thought of G. W. F. Hegel" in Althusser 1997, 78–80. This interpretation includes a series of remarks on key features of the Hegelian account of natural life (such as organism, germ, disease, etc.), which I recall as the chapter develops. On the relationship between Derrida and Althusser, which remains unexplored from the perspective of the Hegelian legacy, see Montag 2013.

5. See Hegel 1999, 145–46: "In the first case, the relation is properly [to be found] in shape and indifference, and the eternal restlessness of the concept, or infinity, lies in part in the organization itself as it consumes itself and relinquishes the appearance of life, the purely quantitative, in order to rise up eternally out of its ashes, as its own seed-corn, to renewed youth."

6. For the French edition of *Reason in History* quoted by Derrida, see Hegel 1965, 76.

7. Cf. Hegel 1965, 77.

8. Therefore, Derrida explains: "And in the description of the spirit that returns to itself through its own proper product, after it lost itself there, there is more than a simple rhetorical convenience in giving to the spirit the name father. Likewise, the advent of the Christian Trinity is more than an empiric event in the spirit's history" (Derrida 1986, 29).

9. See Althusser's remarks on the Hegelian concept of the natural seed and on the relation between natural seed and spiritual nature:

But the resemblance is imperfect, for three reasons: first, because the seed is in externality, drawing sustenance from an earth which is foreign to it; second, because "the seed produced is not identical with the seed from which it came"; finally, because this ovular schema is simple repetition: the seed has no memory, and the content it internalizes is its own past, which, since it repeats itself, is, rather, a present. Nature has no future because it has no history. There is literally nothing new under the sun; the oak is old before it sprouts, whereas the content of Spirit is "always young." Biological circularity is one eternity juxtaposed to another; it reproduces itself because the content is subject to an unmastered necessity. The circularity of the Spirit, in contrast, is a memory that cannot reproduce itself, because it transforms its own law as it gains mastery over it: the developed in-itself produces, not the in-itself pure and simple, but an undecaying totality which absorbs the initial in-itself in its ultimate movement, and does not repeat itself. Spirit is an acorn that produces, not another acorn, but the very tree it fell from a moment ago. This circle is infinite inasmuch as it endlessly completes its own circuit, but it produces itself unaided, and is a circle because the truth is revealed only at the end—when the seed (the in-itself) discovers it is the fruit of the tree which emerges from it. The tree of truth has only ever produced a single seed, the one it sprang from—and the seed learns this only when the tree is full-grown. It is in this sense that the Spirit is Self, that is, auto-development, creation of self by self; and it is in this sense that it is an absolute totality, inasmuch as it is a whole which depends on nothing, and itself posits the origin it springs from. (Althusser 1997, 91–92)

10. For the text of the *Encyclopedia Philosophy of Nature*, see Hegel 1970, 414.

11. Cf. Derrida 1986, 31. See also, for instance, Hegel 1948, 255–56.

12. See also Hegel 1948, 259: "They know God and recognize themselves as children of God, as weaker than he, yet of a like nature in so far as they have become conscious of that spiritual relation suggested by his name (ὄνομα) as the ἄνθρωπος who is φωτιζώμενος φωτῖ ἀληθινῷ [lighted by the true light]. They find their essence in no stranger, but in God."

13. See Derrida 1986, 132–33: "Like every formation, every imposition of form, it is on the male side, here the fathers, and since this violent form bears the parents' death, it matures [*se fait*] above all against the father."

14. The passage continues as follows: "They [the parents] guard in that becoming their own disappearance, reg(u)ard their child as their own death,

they retard it, appropriate it; they maintain in the monumental presence of their seed—in the name—the living sign that they are dead, not the *they are dead*, but that *dead they are*, which is another thing" (Derrida 1986, 133).

15. Derrida dedicates a section of the essay "Qual Quelle: Valery's Sources" (originally presented in 1971, and included in *Margins of Philosophy* in 1972) to the figure of the son as the *Aufhebung* of its genitors (*Der sich aufhebende Ursprung*). In an extended footnote, he explains that the spiritual and human division (*Entzweiung*) of the father into two should be thought in light of the idea, inherited from Jacob Boehme, that negativity and division are at work within the principle of things. It is from this division, almost dissemination, that the source reappropriates itself. Derrida writes:

> The law-of-the-proper, the *economy* of the source: the source is produced only in being cut off (*à se couper*) from itself, only in taking off in its *own* negativity, but equally, and *by the same token*, in reappropriating itself, in order to amortize its own, proper death, to rebound, *se relever*. Reckoning with absolute loss, that is, no longer reckoning, general economy does not cease to pass into the restricted economy of the source in order to permit itself to be encircled. Once more, here, we are reduced to the inexhaustible ruse of the *Aufhebung*, which is unceasingly examined, in these margins, along with Hegel, according to his text, against his text, within his boundary or interior limit: the absolute exterior which no longer permits itself to be internalized. We are led back to the question of dissemination: does semen permit itself to be *relevé*? Does the separation which cuts off the source permit itself to be thought as the *relève* of oneself? And how is what Hegel says of the child to be read in general: "*Der sich aufhebende Ursprung*" (*Realphilosophie of Jena*) or "*Trennung von dem Ursprung*" (*Phenomenology of Spirit*)? (Derrida 1982, 284–85)

16. For a comparative interpretation of the Hegelian tree of life against the Kantian one, see Hamacher 1998, 125–31.

17. Cf. Derrida 1986, 73: ". . . a tree, a vegetable being, a tree of life. The whole circulates in it, from the root toward the top through all the parts. The whole already resides in *le gland* [acorn, glans]."

18. Cf. Hegel 1948, 253. In part I of *The Spirit of Christianity*, on which Derrida had previously commented in *Glas*, Hegel explains that the Jewish family is not yet a family in the Christian sense (which is the truth of family). Abraham, the progenitor of the Jewish nation, did not reconcile with people and nature and thus, as Derrida suggests, he could only found a family (/non-family, cf. Derrida 1986, 36ff.), a genealogical tree, that "takes

root nowhere, never reconciles itself with nature, remains foreign everywhere" (Derrida 1986, 40 and Hegel 1948, 185). For preliminary remarks on metaphoricity and the Jews, see Bennington 2016.

19. Cf. Hegel 1948, 258.
20. Cf. Hegel 1948, 260.
21. Cf. Hegel 1948, 261.
22. For an earlier occurrence of this expression, see Derrida 1978, 95: "Life negates itself in literature only so that it may survive better. So that it may be better. It does not negate itself any more than it affirms itself: it differs from itself, defers itself, and writes itself as *différance*. Books are always books of *life* (the archetype would be the Book of Life kept by the God of the Jews) or of *afterlife* (the archetype would be the Books of the Dead kept by the Egyptians)."
23. Cf. Derrida 1986, 83–84.
24. For a preliminary analysis of preformationist readings of Hegel, see Derrida 1986, 20–21, in which Derrida suggests that these readings would look for preconfiguration and invariance throughout the system.
25. For the idea of a nonsatisfactory metaphor that escapes us, I refer to the remarks that, toward the end of "White Mythology," Derrida dedicates to the ultimate resource of Descartes's metaphorical system—namely, the *lumen natural* that proceeds from God and returns to it. He writes:

> Of course the adoration [of God-light] here is a philosopher's adoration, and since natural light is natural, Descartes does not take his discourse as a theologian's: that is, the discourse of someone who is satisfied with metaphors. And to whom one must leave them: "The author could explain in satisfactory manner, following his philosophy, the creation of the world, such as it is described in Genesis. The narrative of creation found there is perhaps metaphorical; thus, it must be left to the theologians. Why is it said, in effect, that darkness preceded light? . . . And as for the cataracts of the abyss, this is a metaphor, but this metaphor escapes." Presence disappearing in its own radiance, the hidden source of light, of truth, and of meaning, the erasure of the visage of Being—such must be the insistent return of that which subjects metaphysics to metaphor. (Derrida 1982, 267–68)

26. For a reading of the destruction of biological organization from a Hegelian perspective, see Derrida's Artaud in "La Parole Soufflée" (originally published in 1965 and included in *Writing and Difference*, 1967). According to Derrida's reading, Artaud conceives of "organization" as "the membering

and the dismembering of my (body) proper" (Derrida 1978, 234, translation modified). Therefore, the reappropriation of "my body" must go through "the reduction of the organic structure" (235).

27. Cf. Hegel 1970, 410.

28. Cf. Hegel 1970, 415: "The genus particularizes itself, divides itself into its species; and these species, behaving as mutually opposed individuals, are, at the same time, nonorganic nature as the genus against individuality-death by violence."

29. Derrida paraphrases EPN as follows: "The bellicose and morseling operation of the generic process (*Gattungsprozess*) doubles itself with an affirmative reappropriation. The singularity rejoins, repairs, or reconciles itself with itself within the genus. The individual 'continues itself' in another, feels and experiences itself in another" (Derrida 1986, 110).

30. Cf. Hegel 1970, 411. On *Begattung* Derrida writes: "The operation of genus (*Gattung*), the generic and generative operation" (1986, 110).

31. See also Althusser's notes on the Hegelian concept of disease:

> Disease—"an anticipation of death"—represents the animal's supreme effort to possess death even while alive, and to actualize universality for itself. That is why Hegel sees it as the birth of Spirit. But disease is a living contradiction; it can be no more than an anticipated universality, since it is only an anticipated death that persists in being. True universality does not suffer anticipation, which is why the sick individual gets better or dies. In reality, the individual can only attain and possess the universal by tarrying with the universal *in actu*, that is, with death *in actu*. From a natural point of view, man is a living death. (1997, 94)

32. Commenting on this passage, Johnson (1993) explains that Derrida's dissemination also accounts for the non-pure and non-linear *semination* (or insemination) of sexual reproduction, "which, by the mixing of seeds, introduces variety into the chain of being, a variety complicated by the temporal lapse that can occur in the transmission of characters" (162). And he continues: "Unlike the stately return to the same implied in idealist teleology (a single line of descent, a male economy), there is a continuous drift of differences which has no origin and knows of no determinate future. The system is not pulled into the future by a mysterious first (and last) principle, but is pushed *a tergo* by what is handed down, selected and recombined, from its ancestral past. This philosophy is, in essence, a philosophy of evolution" (163).

33. The text is quoted by Derrida in a decisive footnote of the preface to *Dissemination*, in which he observes that Feuerbach turns back against

Hegel the accusations of speculative empiricism and formalism (in Feuerbach's words, of "feint" and "play," Derrida 1981, 40). For a close reading of the aforementioned footnote, see the next chapter.

34. For another version of the same question, see Derrida's aforementioned analysis of *Reason in History*. After recalling the onto-theological figure of the germ and its metaphorical play throughout the regions of the system, Derrida raises the following question: "From where would these figures export themselves? What would be their own proper place?" (Derrida 1986, 27).

35. See Hegel 1970, 417–18: "If we admit that the works of man are sometimes defective, then the works of Nature must contain still more imperfections, for Nature is the Idea in the guise of externality. . . . In Nature, it is external conditions which distort the forms of living creatures; but these conditions produce these effects because life is indeterminate and receives its particular determinations also from these externalities. The forms of Nature, therefore, cannot be brought into an absolute system, and this implies that the species of animals are exposed to contingency."

36. In *Positions* (1972), Derrida observes that Saussure resorts to the concept of classification in order to account for a semiotic code in general ("language and in general every semiotic code—which Saussure define as 'classifications'" Derrida 1981b, 28). Saussure (1959) writes: "Language, on the contrary, is a self-contained whole and a principle of classification. As soon as we give language first place among the facts of speech, we introduce a natural order into a mass that lends itself to no other classification" (9).

CHAPTER 5

1. This text is quoted from Marx's Preface to the first edition of the *Capital* (1867). Derrida's quotation reads: "Every beginning is difficult, holds in all sciences [*dans toutes les sciences le commencement est ardu*]" (Derrida 1981, 34).

2. For a reconstruction of the post-*genetic* critique of the concept of the genetic programme in the early seventies, see Atlan 1979 (especially part 1, 11–131).

3. He calls this logos "inaugural," thus abandoning a phenomenological term that he had employed earlier in order to conceive of the protocollary structure of inscription (cf. my Introduction).

4. The text continues as follows: "Upon reaching the end of the pre- (which presents and precedes, or rather forestalls, the presentative production, and, in order to put before the reader's eyes what is not yet visible, is obliged to speak, predict, and predicate), the route which has been covered must cancel itself out" (Derrida 1981, 9).

5. On Derrida's reading of the programmed suicide of nature as well as of the metaphor and on the hypothesis of a second death prescribed to both, see chapters 3 and 4.

6. I recall that, in the opening page of *Genesis and Structure of Hegel's Phenomenology of Spirit* (1946), Jean Hyppolite dissociates the preface from the introduction as follows: "The preface is an hors d'oeuvre [*hors-d'oeuvre*]; it contains general information on the goal that Hegel sets for himself and on the relation between his work and other philosophical treatises on the same subject. The introduction, on the contrary, is an integral part of the book; it poses and locates the problem, and it determines the means to resolve it" (cf. Hyppolite 1974, 4).

7. For the text commented by Derrida, see Hegel 1977, 1:

For whatever might appropriately be said about philosophy in a preface—say a historical statement of the main drift and the point of view, the general content and results, a string of random assertions and assurances about truth—none of this can be accepted as the way in which to expound philosophical truth. Also, since philosophy moves essentially in the element of universality, which includes within itself the particular, it might seem that here more than in any of the other sciences the subject-matter or thing itself [*die Sache selbst*], even in its complete nature, were expressed in the aim and the final results, the execution [*Ausführung*] being by contrast really the unessential factor [*eigentlich das Unwesentliche sei*].

8. The text from *Science of Logic* section I.2.1.2 continues as follows: "For it is the *unity* of beings which are, only in so far as they are *not one*—and it is the *separation* of beings which are, only in so far as they are separated *in the same reference connecting them*. The positive and the negative, however, are the *posited* contradiction, for, as negative unities, they are precisely their self-positing and therein each the sublating of itself and the positing of its opposite" (Hegel 2010, 374–75).

9. See Derrida 1981, 13: "When the double necessity, both internal and external, will have been fulfilled, the preface, which will in a sense have introduced it as one makes an introduction to the (true) beginning (of the truth), will no doubt have been raised to the status of philosophy, will have been internalized and sublated into it."

10. For Derrida's understanding of "speculative dialectics" as the organization of regional discourses and their sublation in the Hegelian system, see chapter 4.

11. See Derrida 1974, 9:

It is also in this sense that the contemporary biologist speaks of writing and pro-gram in relation to the most elementary processes of information within the living cell. And, finally, whether it has essential limits or not, the entire field covered by the cybernetic program will be the field of writing ... the *grammé*—or the *grapheme*—would thus name the element. An element without simplicity. An element, whether it is understood as the medium or as the irreducible atom, of the arche-synthesis in general, of what one must forbid oneself to define within the system of oppositions of metaphysics, of what consequently one should not even call experience in general, that is to say the origin of meaning in general.

12. For the concept of precipitation, which comes back soon in "Outwork, Prefacing," see Derrida 1978, 20: "It is because writing is inaugural, in the fresh sense of the word, that it is dangerous and anguishing. It does not know where it is going, *no knowledge can keep it from the essential precipitation toward the meaning that it constitutes and that is, primarily, its future* [my emphasis]."

13. On Derrida's reading of this point in relation to Heidegger's introduction to *Being and Time*, see chapter 1.

14. For Feuerbach's text, see Derrida 1981, 29–30:

"Hegelian philosophy presents a contradiction between ... *thought* and *writing*. Formally, the absolute idea is certainly not presupposed, but in essence it is." ... "The estrangement (*Entäußerung*) of the idea is, so to speak, only *pretense*; it makes believe, but it is not in earnest; it is *playing*. The conclusive proof is the beginning of the *Logic*, whose beginning should be the beginning of philosophy in general. Beginning as it does with Being, is a mere formalism, because Being is not the true beginning, the true first term; one could just as easily begin with the absolute Idea, for even before he wrote the Logic, that is, even before he gave his ideas of a scientific form of communication, the absolute Idea was already a certainty for Hegel, an immediate truth." ... "To Hegel the thinker the absolute Idea was an absolute certainty; to Hegel the writer, it was a formal uncertainty."

15. See Derrida 1981, 49: "This is why Hegel never investigates in terms of writing the living circulation of discourse. He never interrogates the exteriority, or the repetitive autonomy, of that textual *remainder* constituted

for example by a preface, even while it is semantically sublated within the encyclopedic logic."

16. In "The Double Session" (first published in 1970 and later included in *Dissemination*), Derrida rejects the understanding of the text as a self-explaining seminal reason by denouncing "a certain imprudence in believing that one could, at last, stop at a textual seed or principle of life referring only to itself" (Derrida 1981, 203). A page prior, he describes the disseminated trace-seed as follows:

> A writing that refers back only to itself carries us *at the same time*, indefinitely and systematically, to some other writing. At the same time: this is what we must account for. A writing that refers only to itself and a writing that refers indefinitely to some other writing might appear noncontradictory: the reflecting screen never captures anything but writing, indefinitely, stopping nowhere, and each reference still confines us within the element of reflection. Of course. But the difficulty arises in the relation between the medium of writing and the determination of each textual unit. It is necessary that while referring each time to another text, to another determinate system, each organism only refer to itself as a determinate structure; a structure that is open and closed *at the same time*. (202)

I remark that here Derrida refers to a reflecting screen of dissemination, the back of the mirror that, as we have just seen, Hegel identifies with nature.

17. Cf. Bachelard 2002, 185–210 (chapter 10).

18. Johnson explains this text by highlighting that the programme of dissemination is situated beyond the limits of the *logos spermatikos*, which also encompass the concept of the so-called genetic programme:

> The code, as Derrida understands it, is therefore constituted in process rather than in anticipation. Despite the suggestion of precedence implied above in Derrida's articulation of the word programme (pro-gramme), the gram, the trace, the inscription are never absolutely primary. There is instead a kind of precipitation towards sense that is ignorant of its future, as was the case with the writing described in "Force et signification" . . . By virtue of a feedback process (both positive and negative) the genetic code is therefore regulating (before) but also regulated (after) in the sense that its pro-gramme is executed in a context that is perpetually changing, hence perpetually modifying the conditions

of possibility of the code. The supplement de code described by Derrida above is this continual differing-from-itself of the code as it descends the evolutionary slope. Of course, the process of selection that operates a posteriori (*après coup*, *Nachträglich*) upon the unprogrammed drift of the code gives the appearance of the necessity of the forms it produces. (Johnson 1993, 169)

For another occurrence of the supplement of code (*supplement de code*), see the final page of "White Mythology": "There is always, absent from every garden, a dried flower in a book; and by virtue of the repetition in which it endlessly puts itself into *abyme*, no language can reduce into itself the structure of an anthology. This supplement of a code which traverses its own field, endlessly displaces its closure, breaks its line, opens its circle, and no ontology will have been able to reduce it" (Derrida 1982, 271).

19. In "Qual Quelle: Valery's Sources," Derrida has recourse to the figure of the self-biting snake in order to account for the problem of auto-insemination. See Derrida 1982, 289: "The circle turns in order to annul the cut, and therefore, by the same token, unwittingly signifies it. The snake bites its tail, from which above all it does not follow that it finally rejoins itself without harm in this successful auto-fellatio of which we have been speaking all along, in truth."

20. As he writes in *Positions*, Derrida seems to demarcate translation from transformation and to prefer the latter for specific reasons:

> But if this difference [between signifier and signified] is never pure, no more so is translation, and for the notion of translation we would have to substitute a notion of transformation: a regulated transformation of one language by another, of one text by another. We will have never, and in fact never have had, to do with some "transport" of pure signifieds from one language to another, or within one and the same language, that the signifying instrument would leave virgin and untouched. (Derrida 1981b, 20)

21. I remark that, here, Wiener (as well as Jacob, who countersigns this expression) describes the relationship between writing and the living as metaphorical. Conversely, as pointed out in the aforementioned footnote on dissemination and the philosophy of the seed, Derrida is seeking a more than metaphorical articulation of writing and biological genesis—that is, the trace-seed as the general structure of genesis.

22. It is worth recalling that Derrida engages in a more explicit examination of the philosophical underpinnings of Jacob's biological theory and, more generally, of the biological thought of his time, in the unedited

seminar entitled *La Vie la mort* (1975), where he offers a close reading of *The Logic of Life*. For the exploration of this text, which goes beyond the purpose of my book, I refer to Francesco Vitale's forthcoming work entitled *Biodeconstruction*.

POSTSCRIPT

1. The exergue is a quotation from Derrida's essay "Divided Bodies: Response to *Nouvelle Critique*" (originally published in 1975).
2. For another reading of this essay, see Wortham 2006, 68–84.
3. On this point, see Derrida 2002b, 131.
4. See Derrida 2002b, 79:

> Consequently, when I say, in such a trivial formula, that power controls the teaching apparatus, it is not to place power outside the pedagogic scene. (Power is constituted inside pedagogy as an effect of this scene itself, no matter what the political or ideological nature of the power in place around it.) Nor is it to make us think or dream of a teaching without power, free from teaching's own power effects or liberated from all power outside of or higher than itself. That would be an idealist or liberalist representation, with which a teaching body blind to power—the power it is subject to, the power at its disposal in the place where it denounces power—effectively reinforces itself.

5. For the understanding of *différance* in relation to differences of forces, see the following passage from "Différance" (originally published in 1968 and then included in *Margins of Philosophy*, 1972), where Derrida inscribes this understanding within a Nietzschean and Freudian tradition:

> Thus, *différance* is the name we might give to the "active," moving discord of different forces, and of differences of forces, that Nietzsche sets up against the entire system of metaphysical grammar, wherever this system governs culture, philosophy, and science. It is historically significant that this diaphoristics, which, as an energetics or economics of forces, commits itself to putting into question the primacy of presence as consciousness, is also the major motif of Freud's thought: another diaphoristics, which in its entirety is both a theory of the figure (or of the trace) and an energetics. The putting into question of the authority of consciousness is first and always differential. (Derrida 1982, 18)

6. For Derrida's later elaboration of this text, I recall the following passage from *Specters of Marx*:

> It is at this point that Marx intends to distinguish between the spirit (*Geist*) of the revolution and its specter (*Gespenst*), as if the former did not already call up the latter, as if everything, and Marx all the same recognizes this himself, did not pass by way of differences *within a fantastics as general as it is irreducible*. Far from organizing the good schematics of a constitution of time, this other transcendental imagination is the law of an invincible *anachrony*. (Derrida 1994, 140)

7. Cf. Derrida 1975, 14.9–10.
8. For the entry on the drive for mastery, see Laplanche and Pontalis 1973, 217–18.
9. Cf. Derrida 1987, 325 and 403–4.

BIBLIOGRAPHY

Althusser, Louis. 1997. *The Spectre of Hegel: Early Writings*. Translated by Geoffrey Michael Goshgarian. London: Verso.
———. 2003. *The Humanist Controversy and Other Writings (1966–67)*. Translated by Geoffrey Michael Goshgarian. London: Verso.
Aristotle. 1984. *Complete Works (I–II): The Revised Oxford Translation*. Edited by Jonathan Barnes. Princeton, NJ: Princeton University Press.
Atlan, Henry. 1972. *L'Organisation biologique et la théorie de l'information*. Paris: Hermann.
———. 1979. *Entre le cristal et la fumée. Essai sur l'organisation du vivant*. Paris: Seuil.
———. 2011. *Selected Writings: On Self-Organization, Philosophy, Bioethics, and Judaism*. Edited by Stefanos Geroulanos and Todd Meyers. New York: Fordham University Press.
Bachelard, Gaston. 2002. *The Formation of the Scientific Mind*. Translated by Mary McAllester Jones. Manchester, UK: Clinamen Press.
Baring, Edward. 2011. *The Young Derrida and French Philosophy, 1945–1968*. Cambridge, UK: Cambridge University Press.
Bennington, Geoffrey. 2010. *Not Half No End: Militantly Melancholic Essays in Memory of Jacques Derrida*. Edinburgh: Edinburgh University Press.
———. 2014. "Metaphor and Analogy in Derrida." In *A Companion to Derrida*, edited by Zeynep Direk and Leonard Lawlor, 89–104. Chichester, UK: John Wiley and Sons.
———. 2016. "Notes towards a Discussion of Method and Metaphor in Glas." *Paragraph* 39:2, 249–64. Edinburgh: Edinburgh University Press.
———, and Jacques Derrida. 1993. *Jacques Derrida*. Chicago: University of Chicago Press.
Bourgeois, Bernard. 1969. *Hegel à Francfort. Judaïsme, christianisme, hégélianisme*. Paris: Vrin.
Bouton, Christophe. 2000. *Temps et esprit dans la philosophie de Hegel: De Francfort à Iéna*. Paris: Vrin.

———, and J.-L. Vieillard-Baron, eds. 2009. *Hegel et la philosophie de la nature*. Paris: Vrin.

Brisson, Luc. 1974. *Le même et l'autre dans la structure ontologique du Timée de Platon. Un commentaire systématique du Timée de Platon*. Paris: Klincksieck.

Cahiers de Royaumont. 1965. *Le concept d'information dans la science contemporaine*. Paris: Les Editions de Minuit.

Canguilhem, Georges. 1989. "Vie." In *Encyclopaedia Universalis* 23: 546–53.

Cassin, Barbara, ed. 2014. *Dictionary of Untranslatables: A Philosophical Lexicon*. Translation edited by Emily Apter, Jacques Lezra, and Michael Wood. Princeton, NJ: Princeton University Press.

Derrida, Jacques. 1969–1970. *Théorie du discours philosophique: La métaphore dans le texte philosophique*. Caen, France: IMEC.

———. 1970–1971. *Théorie du discours philosophique: Conditions de l'inscription du texte de philosophie politique—l'exemple du matérialisme*. Caen, France: IMEC.

———. 1973. *Speech and Phenomena and Other Essays on Husserl's Theory of Signs*. Translated by David Allison and Newton Garver. Evanston, IL: Northwestern University Press.

———. 1974. *Of Grammatology*. Translated by Gayatri Chakravorty Spivak. Baltimore, MA: Johns Hopkins University Press.

———. 1974–1975. *GREPH (Le concept de l'idéologie chez les idéologues français)*. Caen, France: IMEC.

———. 1975. *La Vie la mort*. Caen, France: IMEC.

———. 1978. *Writing and Difference*. Translated by Alan Bass. London: Routledge and Keagan Paul Ltd.

———. 1981. *Dissemination*. Translated by Barbara Johnson. Chicago: University of Chicago Press.

———. 1981b. *Positions*. Translated by Alan Bass. Chicago: University of Chicago Press.

———. 1982. *Margins of Philosophy*. Translated by Alan Bass. Brighton, UK: Harvester Press.

———. 1985–1986. *Littérature et philosophie comparées: Nationalité et nationalisme philosophique: Mythos, logos, topos*. Caen, France: IMEC.

———. 1986. *Glas*. Translated by John P. Leavey Jr. and R. Rand. Lincoln, NE: University of Nebraska Press.

———. 1987. *The Postcard. From Socrates to Freud and Beyond*. Translated by Alan Bass. Chicago: University of Chicago Press.

———. 1989. *Edmund Husserl's* Origin of Geometry: *An Introduction*. Translated by John P. Leavey. Lincoln, NE: University of Nebraska Press.

———. 1994. *Specters of Marx: The State of the Debt, the Work of Mourning and the New International*. Translated by Peggy Kamuf. New York: Routledge.

———. 1995. *On the Name*. Translated by David Wood, John P. Leavey, and Ian McLeod. Stanford, CA: Stanford University Press.

———. 1997. *The Politics of Friendship*. Translated by George Collins. London: Verso.

———. 2002. *Acts of Religion*. Edited by Gil Anidjar. London: Routledge.

———. 2002b. *Who Is Afraid of Philosophy: Right to Philosophy I*. Translated by Jan Plug. Stanford, CA: Stanford University Press.

———. 2005. *Rogues. Two Essays on Reason*. Translated by Pascale-Anne Brault and Michael Naas. Stanford, CA: Stanford University Press.

———. 2005b. "A Time for Farewells: Heidegger (read by) Hegel (read by) Malabou." Translated by Joseph Cohen. In Catherine Malabou, *The Future of Hegel: Plasticity, Temporality and Dialectic*. Translated by Lisbeth During, vii–xlvii. London: Routledge.

———. 2016. *Heidegger: The Question of Being and History*. Translated by Geoffrey Bennington. Chicago: University of Chicago Press.

Ferrini, Cinzia. 2011. "Hegel on Nature and Spirit: Some Systematic Remarks." *Hegel-Studien* 46: 117–50. Hamburg: Felix Meiner Verlag.

Feuerbach, Ludwig. 1960. *Manifestes philosophiques: textes choisis (1839–1845)*. Edited by Louis Althusser. Paris: Presses Universitaires de France.

Freud, Sigmund. 1953–1974. *The Standard Edition of the Complete Psychological Works of Sigmund Freud, Volume XX (1925–1926)*. Edited by James Strachey. London: Hogarth Press.

Gasché, Rodolphe. 1986. *The Tain of the Mirror. Derrida and the Philosophy of Reflection*. Cambridge, MA: Harvard University Press.

———. 1994. *Inventions of Difference*. Cambridge, MA: Harvard University Press.

———. 2014. "The Eve of Philosophy: On 'Tropic' Movements and Syntactic Resistance in Derrida's *White Mythology*." In *International Yearbook for Hermeneutics* 13:1–22. Tübingen, Germany: Mohr Siebeck.

Gratton, Peter. 2015. "The Spirit of the Time: Derrida's Reading of Hegel in the 1964–65 Lecture Course." *CR: The New Centennial Review* 15.1: 49–65. East Lansing: Michigan State University Press.

Hamacher, Werner. 1998. *Pleroma: Reading in Hegel*. Translated by Nicholas Walker and Simon Jarvis. Stanford, CA: Stanford University Press.

Hegel, Georg Wilhelm Friedrich. 1941. *Phénoménologie de l'esprit*. Translated by Jean Hyppolite. Paris: Aubier, Editions Montaignes.

———. 1948. *Early Theological Writings*. Translated by Thomas Malcom Knox. Chicago: University of Chicago Press.

———. 1965. *La Raison dans l'histoire*. Translated by Kostas Papaioannou. Paris: Union générale d'édition.

———. 1970. *Philosophy of Nature*. Translated by Arnold V. Miller. Oxford, UK: Oxford University Press.

———. 1977. *Phenomenology of Spirit*. Translated by Arnold V. Miller. Oxford, UK: Oxford University Press.

———. 1979. *System of Ethical Life and First Philosophy of Spirit*. Translated by Henry S. Harris and Thomas Malcom Knox. Albany, NY: State University of New York Press.

———. 1987. *Jenaer Systementwürfe III. Naturphilosophie und Philosophie des Geistes*. Neu hrsg. Rolf–Peter Horstmann. Hamburg: Felix Meiner Verlag.

———. 1999. *Political Writings*. Translated by Hugh Barr Nisbet. Cambridge, UK: Cambridge University Press.

———. 2010. *Science of Logic*. Translated by George Di Giovanni. Cambridge, UK: Cambridge University Press.

———. 2010b. *Encyclopedia of the Philosophical Sciences in Basic Outline (Part I: Science of Logic)*. Translated by Klaus Brinkmann and Daniel O. Dahlstrom. Cambridge, UK: Cambridge University Press.

Heidegger, Martin. 1996. *Being and Time*. Translated by Joan Stambaugh. Albany, NY: State University of New York Press.

Hyppolite, Jean. 1974. *Genesis and Structure of Hegel's Phenomenology of Spirit*. Translated by Samuel Cherniak and John Heckman. Evanston, IL: Northwestern University Press.

———. 1991. *Figures de la pensée philosophique. Ecrits 1931–1968*. Paris: Presses universitaires de France.

Illetterati, Luca. 1985. *Natura e Ragione. Sullo sviluppo dell'idea di natura in Hegel*. Trento: Verifiche.

Jacob, François. 1973. *The Logic of Life: A History of Heredity*. Translated by Betty Spillmann. New York: Pantheon Books.

Johnson, Christopher. 1993. *System and Writing in the Philosophy of Jacques Derrida*. Cambridge, UK: Cambridge University Press.

Laplanche, Jean, and J.-B. Pontalis. 1973. *The Language of Psychoanalysis*. Translated by Donald Nicholson–Smith. London: Hogarth Press.

Lawlor, Leonard. 2002. *Derrida and Husserl: The Basic Problem of Phenomenology*. Bloomington, IN: Indiana University Press.

Leibniz, G. W. 1952. *Theodicy. Essays on the Goodness of God, the Freedom of Man and the Origin of Evil*. Translated by E. M. Huggard. New Haven, CO: Yale University Press.

———. 1989. *Philosophical Papers and Letters*. Edited by Leroy M. Loemker. Dordrecht: Kluwer Academic Publishers.

———. 1996. *New Essays on Human Understanding*. Edited by Peter Remnant and Johnathan Bennett. Glasgow, UK: Cambridge University Press.

Loraux, Nicole. 1993. *The Children of Athena: Athenian Ideas about Citizenship and the Division between the Sexes*. Translated by Caroline Levine. London: Routledge.

Malebranche, Nicolas. 1997. *The Search after Truth*. Translated by Thomas M. Lennon and Paul J. Olscamp. Cambridge: Oxford University Press.

Marrati, Paola. 2005. *Genesis and Trace: Derrida Reading Husserl and Heidegger.* Stanford, CA: Stanford University Press.

Marx, Karl. 1972. *The German Ideology.* Edited by Christopher J. Arthur. New York: International Publishers.

———. 1976. *Capital: A Critique of Political Economy (I).* Translated by Ben Fowkes. London: Penguin Classics in association with New Left Review.

Margel, Serge. 1995. "Les nourritures de l'âme. Essai sur la fonction nutritive et séminale dans la biologie d'Aristote." *Revue des Études Grecques* 108 (January–June): 91–106.

Marmasse, Gilles. 2008. *Penser le réel: Hegel, la nature et l'esprit.* Paris: Editions Kimé.

Migliori, Maurizio. 2003. "Il problema della generazione nel Timeo." In *Plato Physicus, Cosmologia e antropologia nel Timeo*, 97–120. Amsterdam: Adolf M. Hakkert.

Montag, Warren. 2013. *Althusser and His Contemporaries: Philosophy's Perpetual War.* Durham, NC: Duke University Press.

Naas, Michael. 2010. "Earmarks: Derrida's Reinvention of Philosophical Writing in 'Plato's Pharmacy.'" In *Derrida and Antiquity*, edited by Miriam Leonard, 35–57. Oxford: Oxford University Press.

———. 2014. "Derrida and Ancient Philosophy (Plato and Aristotle)." In *A Companion to Derrida*, edited by Zeynep Direk and Leonard Lawlor, 231–50. Chichester, UK: John Wiley and Sons.

Nancy, Jean-Luc. 2001. *The Speculative Remark: One of Hegel's Bon Mots.* Translated by Céline Surprenant. Stanford, CA: Stanford University Press.

Plato. 1997. *Complete Works.* Edited by John M. Cooper. Indianapolis, IN: Hackett.

———. 2003. *Timeo.* Translated by Francesco Fronterotta. Milano: Biblioteca Universale Rizzoli.

———. 2007. *Sofista.* Translated by Francesco Fronterotta. Milano: Biblioteca Universale Rizzoli.

Protevi, John. 1994. *Time and Exteriority: Aristotle, Heidegger, Derrida.* London: Associated University Press

Roger, Jacques. 1998. *The Life Sciences in Eighteenth–Century French Thought.* Translated by Robert Ellrich. Stanford, CA: Stanford University Press.

Rousset, Jean. 1962. *Forme et signification. Essai sur les structures littéraires de Corneille à Claudel.* Paris: José Corti.

Sallis, John. 1999. *Chorology: On Beginning in Plato's Timaeus.* Bloomington, IN: Indiana University Press.

Saussure, Ferdinand de. 1959. *Course in General Linguistics.* Translated by Wade Baskin. New York: Philosophical Library.

Senatore, Mauro. 2015. "Of Natural Metaphors. Derrida on the Eluded Necessity of the Hegelian System." *Revista Eletrônica Estudos Hegelianos*, 12, 20: 71–93.

Stone, Alison. 2005. *Petrified Intelligence: Nature in Hegel's Philosophy*. Albany, NY: State University of New York Press.
Vitale, Francesco. 2013. "The Law of the Oikos. Jacques Derrida and the Deconstruction of the Dwelling." *SAJ: Serbian Architectural Journal* 5.1: 59–74.
———. 2014. "The Text and the Living: Jacques Derrida between Biology and Deconstruction." *The Oxford Literary Review* 36.1: 95–114. Edinburgh: Edinburgh University Press.
———. 2015. "Life, Death and Difference: Philosophies of Life between Hegel and Derrida." *CR: The New Centennial Review* 15.1: 93–112. East Lansing: Michigan State University.
Wortham, Simon Morgan. 2006. *Counter-Institutions: Jacques Derrida and the Question of the University*. New York: Fordham University Press.

INDEX

abyss, 13, 57, 66–67, 101, 161n26, 168n25, 173n18
Althusser, Louis, 79, 148n5, 149n7, 165n3, 165n9, 169n31
anagram, 26, 36–38, 46–48, 50, 69, 78–79, 89–90, 92, 143, 154n15, 155n21, 156n31
analogy, 5, 43, 47, 58–62, 83, 93–98, 100, 102–4, 110, 123, 129–30, 156n31, 159n15, 164n1
anguish, 3–5, 7, 11, 148n4, 149n8, 150n11, 172n12
angustia. *See* anguish
animal, the, 20, 33, 44, 53, 91, 94–100, 108–13, 136, 153n12, 169n31, 170n35
archē, 13, 30–31, 87, 171n11
Aristotle: actuality (potentiality), 2, 20–23, 152n27, 160n21; homonymy, 162n9; matter, 63; movement, 22–23; *philosophia prōtē*, 70, 74, 88, 161n1; teleology, 107
Atlan, Henri, xi–xii, 116, 147n2, 170n2

Bachelard, Gaston, 132, 135, 173n17
Bachelard, Suzanne, 79
Bataille, George, 85–86
beginning, 16–17, 20, 31, 51–52, 58, 79–80, 83, 95–97, 104–6, 115–16, 118, 123, 125–28, 130, 140–41, 158n9, 170n1, 171n9, 172n14
Bernard, Claude, 136
Boehme, Jacob, 167n15
book of life, xiii, 19, 23, 69–70, 82, 93, 105, 114–15, 118–20, 124, 168n22
Bourgeois, Bernard, 105–6

Canguilhem, Georges, 147n3
classification, 94, 111–14, 130, 132, 149n8, 170n36
code, 116, 133–37, 170n36, 173n18
concatenation, 26–27, 31, 35–38, 46–47, 155n23
cybernetics, 15, 124, 134, 137, 171n11

Derrida, Jacques: "The Crisis in the Teaching of Philosophy," 143; "Différance," 175n5; "The Double Session," 164n22, 173n16; *Edmund Husserl's* Origin of Geometry: *An Introduction*, xii, 8–11, 78–80, 89, 126, 150n11, 150n13, 151n17, 151n19; "Force and Signification," 1–12, 14–20, 22–23, 120–21, 124–26, 132–33, 149nn9–10, 154n14, 172n12; "Force of Law," 148n4; "Freud and the Scene

Derrida, Jacques *(continued)*
of Writing," 149n8; "From
Restricted to General Economy:
A Hegelianism without Reserve,"
85–87; *Glas*, 12–13, 70, 80, 84,
87, 91, 93–115, 131, 149n8,
154n13, 163n15, 163n18, 164n1,
165n7, 166n11, 166nn13–14,
167nn17–18, 168n24, 169nn29–
30, 170n34; *GREPH (Le concept
de l'idéologie chez les idéologues
français)*, 144–45; *Heidegger: The
Question of Being and History*, 25–
31, 82–86, 153n5, 153n8, 157n1;
Khōra, 50–51, 57, 60–61, 63–68,
160n25; *Of Grammatology*, 14–15,
37, 113–14, 124, 132–33, 150n15,
151n20, 155n21, 171n11; "*Ousia*
and *Grammē*," 152n26; "Outwork,
Prefacing [*Hors livre, préfaces*],"
115–36, 151n21, 169n33, 170n4,
171n9, 172nn14–15; *Littérature et
philosophie comparées: Nationalité et
nationalisme philosophique: Mythos,
logos, topos*, 50–51, 53, 55, 61–62,
157n4, 158n10, 160nn17–21;
"Parole Soufflée, La," 168n26;
"Philosophy and its Classes,"
143; "Plato's Pharmacy," 25–26,
28–29, 31–50, 56–57, 71–73,
89, 153nn9–12, 154–55nn15–19,
156nn24–28, 156–n30, 156n32,
157n33, 163n20; *The Politics
of Friendship*, 157n4; *Positions*,
170n36, 174n20; "Qual Quelle:
Valery's Sources," 167n15, 174n19;
Rogues. Two Essays on Reason, 49,
160n22; *Specters of Marx*, 161n4,
176n6; "To Speculate—On
'Freud,'" 145, 176n9; *La théorie de
la signification dans les* Recherches
logiques *et dans* Ideen I, 162n12;
*Théorie du discours philosophique:
Conditions de l'inscription du texte
de philosophie politique l'exemple
du matérialisme*, 50, 57–61, 158n8,
159nn13–14; *Théorie du discours
philosophique: La métaphore dans
le texte philosophique*, 70–76,
79–82, 84–85, 88, 161n1, 161n4,
161n6, 162n8, 162n13; *La Vie
la mort*, 145, 174n22; "Violence
and Metaphysics," 74–75, 77–78,
155n20; "Where a Teaching
Body Begins and how It
Ends," 140–43, 175n4; "White
Mythology: Metaphor in the Text
of Philosophy," 70, 78, 80, 87–92,
155n23, 162n9, 163n19, 164n22,
168n25, 173n18

differance. *See* seed: seminal
differance

dissemination, xii, xiii, 1, 23, 25,
42–44, 48, 50, 58, 61, 72, 74, 82,
92, 100, 111–33, 135, 138–39,
142, 144–45, 164n22, 167n15,
169n32, 173n16, 173n18, 174n21.
See also seed: trace-seed

equivocity, 3–4, 11–12, 71–72,
76–81, 84, 86–90, 109, 124–25,
128–29, 148n3, 150n11, 155n22,
162n13

family, 30, 34, 42–44, 71–72,
95–100, 102, 106, 113, 149n8,
154n13, 161n2, 167n18
father: father of the logos, xiii,
25, 31–40, 42–44, 49, 71–73,
116, 129, 153nn10–11, 154n15;
fatherland, the, 51–57, 61, 83,
157n4; father-son, 57–60, 67–68,
90–92, 95–106, 149n8, 165n7,
166n13, 167n15

INDEX

Feuerbach, Ludwig, 6–7, 79–80, 112, 126–27, 148n5, 149n7, 169n33, 172n14
Fink, Eugen, 79, 150n16
force, 3, 12–13, 15–16, 38, 55, 76, 82–83, 107, 136, 140–45, 155n19, 156n32, 163n16, 175nn4–5
Freud, Sigmund, 4–5, 133, 145, 175n5

general text, 115, 120, 124–37
genetic programme. *See* genetics
geneticism, 2, 15–16, 19–23, 151n21
genetics, xi–xiii, 2, 16, 115–17, 124, 132–38, 147n4, 151n21, 170n2, 173n18
germ, 17, 19–21, 91, 94–100, 110–12, 114, 120, 123–24, 128–30, 132–34, 138–39, 144, 149n8, 165n3, 170n34
GREPH, 139, 141, 144, 147n5

Hegel, G. W. F.: *Aufhebung*, 6–7, 74, 78, 80 81, 84–88, 99–100, 109–10, 112, 148n6, 162n7, 162n13, 167n15; bond, 101–2; contradiction, 74–75, 80–81, 85, 94, 109–13, 119–20, 142, 145, 169n31, 171n8, 172n14; disease, 107–8, 110–11, 130, 165n3, 169n31; education, 90–91, 99–100; exposition, 18–19, 118–19, 124, 126, 128–29; freedom, 85–86, 94–96, 165n2; genus, 6–7, 98, 107–13, 148n6, 169nn28–30; German language, 37, 74–78, 80, 155n22, 163n18; impotence, 94, 97, 101–2, 112, 126; introduction, 115–16, 123–24, 128, 135, 171n6, 171n9; the Jews, 101–6, 167n18, 168n22; philosophy of nature, 2, 6–7, 14, 71, 81–84, 91–92, 98, 101, 106–7, 128–31, 144, 148n6, 163n14, 165n3; phoenix, 94–95, 101, 111–12; *Potenz*, 12–14, 99; propitious time, 121; Proteus, 131; sciences, 72–75, 80–81, 83–87, 90, 119, 123, 128, 161n4, 171n7; sex–relation, 98, 108–10, 169n32; sexual difference, 94, 99, 108–13, 154n13; *Sinn*, 75, 78, 89, 124–25; the tree, 82–83, 101–6, 166n9, 167nn16–18
Heidegger, Martin: destruction, 26–27, 30, 82, 153n1; *Sophist*, interpretation of the, 28–29, 56–59, 61–62, 153nn5–6, 157n1; telling stories, 25, 28–32, 35–36, 40, 89–90, 153n5
history: concept of history (historicity), 15–16, 22–23, 50, 61–68, 79, 82, 132, 143, 149n7, 154n16, 165n7; historical genesis, 9–12, 14, 19, 82–83, 115, 118, 124–26, 131, 133, 150n13; history of Athens, 51–52, 55–56; history of consciousness, 17–19; history of language, 81, 89–90, 155n23, 162n8; history of ontology, 26, 29–30, 47; history of the interpretations of *khōra*, 60–61, 63–66; history of the life sciences, 136–38; nonhistoricity, 98, 110, 165n9; prehistory, 11–12, 151n19. *See also* Heidegger: telling stories
Husserl, Edmund: consciousness, 74; Kant, 10–11, 150n16; origin of geometry, xii, 1, 8–12, 120–21, 151n17; transcendental language, 9–11, 78–79, 89, 150n11, 151n19
Hyppolite, Jean, 152n23, 171n6

inscription: genetic inscription, xi–xii, 1–5, 7–8, 10, 14–15, 22,

inscription *(continued)*
  36–37, 46, 85, 89, 113–14, 120, 126–28, 130–35, 139, 142, 149n8, 149n10, 150n10, 151n19, 154n15, 170n3, 173n18; in the *matrix*, 45, 50, 61–62, 66–67, 122, 160n25; in the soul, 38–39, 42–43, 45; reinscription, 11–12, 37, 85, 126–27, 150n11

Jacob, François, xiii, 116, 134–38, 147n1, 174nn21–22

khōra, 41–42, 45, 49–69, 89, 122, 143–44, 153n6, 157–61nn1–27

lag, 121–22
Leibniz, G. W.: philosophical language, 77, 162n11; scene of creation, 1–5, 7–12, 14–15, 19–20, 116–17, 138, 147n1, 148nn2–3, 149n10
logos spermatikos, xii, 2, 16, 19–21, 23, 35, 45–46, 56–57, 69, 71–73, 79, 84, 86, 120, 132, 135, 137, 140, 154n14, 173n18
logos-*zōon*, xiii, 25, 33–37, 40–47, 49, 54, 58, 61–62, 67–69, 71–74, 80, 117
Loraux, Nicole, 41–42, 51–57, 61, 157n3, 158nn6–7
Lwoff, André, 15

Malebranche, Nicolas, 21–22
Marx, Karl, 57, 128, 140, 144–45, 176n6
mastery, xiii, 4–5, 38, 71–72, 84–88, 92, 116–17, 125–26, 131, 139–40, 143–45, 155n19, 156n28, 165n9, 176n8
metaphor: death of metaphor, 91–92, 118, 171n5; graphic metaphor, 42, 113, 135, 174n21; metaphor of life, 102, 105, 128; metaphorical play, xiii, 34–35, 37, 43, 69–70, 73, 93, 95–97, 101–2, 107, 119–20, 129, 155n23, 170n34; metaphoricity, 35, 81, 102–3, 130, 164n22, 167n18; concept of metaphor (metaphorization), 71, 87–90; ontic metaphor, 27, 31; metaphors, 50, 58–59, 90–91, 106, 155n13, 168n25; natural image (metaphor), 17, 33–34, 82, 93–95, 97, 101–2, 104–5, 111–12, 129, 163n15; nonmetaphorical relationship, xiii, 25, 31, 33, 35, 42, 47, 49, 52, 54, 68, 71, 73, 154n15; more than metaphorical, 133, 174n21; onto-theological figure, 95, 100, 102, 104, 170n34
mise en abyme. *See* abyss
modernity, 1, 9, 92, 115, 118, 124–25, 136, 143, 152n22
mother, 5, 41, 45, 50–60, 62, 64, 67–68, 73–74, 157n4
mythology, 25–26, 29–31, 35–36, 40–41, 47, 54–57, 64, 79–80, 126, 153n5, 153n10, 154n16, 157n1, 158n6

Novalis, 131–32

ontology, xiii, 26–31, 46–47, 59, 61–62, 69, 93, 99, 112, 118–19, 124, 128, 140–41, 159n13, 173n18
onto-theology, 28, 95–96, 98, 100–1, 104, 113–14, 120, 129, 132, 140–41, 170n34

Plato: autochthony, xiii, 39, 41–42, 49, 51–57, 61, 67–68, 157n4, 158n7, 160n17; dialectic, 33, 39, 42, 45, 47–48, 59, 61–62, 159n14; *dynamis*, 61, 63, 160n18,

INDEX

160nn20–21; *eugeneia* (noble birth), 25, 33–34, 40–43, 45, 52, 56, 153n11, 157n4; grammar, 26, 35–38, 46–48; memory, 31–32, 36, 38–39, 42, 56; origin of the world, 26, 30, 45–46, 48–50, 58, 67; pederasty, 43–44; *pharmakon*, 31–32, 36–38, 40, 89; *pharmakos*, 40–43, 156n25, 156n28; receptacle, 45, 50, 59–65, 68, 159n15, 160n17; *symplokē* (*sumplokē*), 47–48, 59, 61, 157n33, 157n35; *triton genos* (third term), 47–49, 57–59, 61–63, 67, 157n1, 158n11, 159nn13–14; Theuth, 26, 31–32, 35–36, 38–39, 46. *See also* analogy; dissemination; Heidegger: *Sophist*; khōra; father; logos-*zōon*; mother; seed

power, xiii, 3–4, 12–13, 21, 27, 31–36, 42–43, 71, 110–11, 120, 139–45, 156n24, 160n18, 160n20, 175n4

precipitation, 11, 14, 124–25, 127, 132–33, 172n12, 173n18

preformationism, xii, 2, 15–16, 19–21, 23, 84, 105, 137–38, 154n14, 168n24

protocol, 116–18, 126–27, 135, 155n20, 170n3

Proust, Marcel, 16–20

regulation, 15, 66–67, 71–72, 93–94, 98, 102, 120, 123–24, 173n18, 174n20

remark, 76, 93, 95–96, 98–101, 104–5, 109, 112

Roger, Jacques, 2, 20–22

Rousset, Jean, 2, 15–19, 22

Schelling, Friedrich, 12–13

seed: human seed, 152n27; natural seed, xii, 1–2, 6–7, 14, 20–21, 43–45, 82–83, 96–97, 101, 125, 149n8, 156n32, 165n9, 169n32; philosophy of the seed, 132, 135, 138, 174n21; seminal differance, 129–30, 132–33, 135; seminal reason, 21, 132–36, 173n16; spiritual seed, 7, 99–101, 129, 165n24, 166n14; trace-seed, xii–xiii, 2, 8, 44–45, 48, 50, 58, 61, 67, 69, 72, 91, 113–14, 117, 120, 125, 131, 150n10, 157n32, 164n22, 173n16, 174n21

self-reproduction, xi–xii, 2, 6–7, 17, 19, 21, 56, 68, 73–74, 86–87, 90, 96–102, 120, 129, 132, 165n9

simultaneity, 16, 19–21, 103, 132, 138, 162n13

speculative, 14–15, 37, 74–78, 80–81, 85, 87, 89, 96, 99, 103, 123–25, 127–28, 154n13, 155n22, 161n6, 162n7, 163n18, 169n33, 171n10

syllogism, 98, 105, 128–31

syntax, 26–27, 37, 46–47, 71–72, 77–79, 89–90, 92, 120, 122, 125, 133–34, 155n20, 155n23, 164n22

teaching body, 139–44, 175n4

textuality, 35–38, 40, 46–48, 50, 69, 79, 89–90, 115, 127, 133–34

trace, 22, 26, 43, 45–46, 48, 50, 59, 76, 103, 149n8, 150n11, 156n32, 173n18, 175n5. *See also* seed: trace-seed

tropic. *See* tropology

tropology, 25–26, 59–61, 63–66, 78–79, 87–92, 120, 122, 125, 133–34, 143, 155n23

Vernant, Jean-Pierre, 50, 64

Wiener, Norbert, 134–35, 174n21

www.ingramcontent.com/pod-product-compliance
Lightning Source LLC
Chambersburg PA
CBHW030654230426
43665CB00011B/1088